The
River Dragon
Has Come!

Chapter One

The Three Gorges Project

A Symbol of Uncontrolled Development in the Late Twentieth Century

Dai Qing

"Water benefits all things generously and without strife. It dwells in the lowly places that men disdain. Thus it comes near to the Dao."

—Laozi

The opening of my country to the outside world has been the most important development in twentieth century China. The two major consequences of this "opening" have been the birth, development, and dominance of the communist/socialist system, and the influx of modern science and technology. We Chinese are repeatedly told that both the communist system and the ascendency of science and technology fit China's historical conditions of economic underdevelopment, foreign domination, and political autocracy. But rather than "fit" our national conditions, these systems have dominated and distorted our lives. As the old Chinese adage says: "Things will develop in the opposite direction when they become extreme" (*wuji bifan*). This is the case with our current socialist regime and its blind faith that engineers and technical fixes can solve all problems. The result of all this is uncontrolled development, and there is no better symbol of uncontrolled development than the Three Gorges dam.

"Uncontrolled" (*bujia jiezhi*) and "out of control" (*shiqu kongzhi*) are similar terms which actually have different implications: The first—uncontrolled—is subjective and describes someone who consciously fails to control his/her behavior. The second—out of control—is more objective and describes how someone's behavior can cause things to spin out of control.

3

The Three Gorges project has been meticulously planned and controlled from its original design to its final construction. But the people who have been doing this planning have failed to understand key Chinese concepts such as self-restraint and the control of brazen arrogance. In Chinese antiquity, a sense of self-restraint was paramount; as the ancient Daoist philosopher Laozi said: "To know one's limits is to be invincible" (*zhizhi keyi budai*). But a couple of centuries after the advent of the industrial revolution, this ancient wisdom lost its appeal and has only been recalled in the last fifty years. This conscious failure by China's leaders to "control" their behavior; that is, to respect and follow ancient wisdom, is what makes the Three Gorges dam a symbol of uncontrolled development. The sad irony is that although every aspect of the Three Gorges dam's construction has been thoroughly planned by scientists, engineers, and officials, if it is completed and goes into operation, we will quickly learn that we are unable to control its effects on the environment, and on society.

The Three Gorges dam will be the largest dam ever built. Its wall of concrete, reaching 185 meters into the air and stretching almost two kilometers across, will create a 600–kilometer-long reservoir.

The dam will require technology of unprecedented sophistication and complexity: It will include twenty-six, 680 MW turbines; twin five-stage lock systems, and the world's highest vertical shiplift.

The project will also cause some of the most egregious environmental and social effects ever: It will flood 30,000 hectares of prime agricultural land in a country where land is the most valuable resource; it will cause the forcible resettlement of upward of 1.9 million people; it will forever destroy countless cultural antiquities and historical sites; and it will further threaten many endangered species, some already facing extinction.*

But perhaps the most astounding fact of all is that although the project has attracted the interest of the world's businesses and the ire of its environmentalists, it has faced very little opposition at home. The National People's Congress (NPC) approved the project in April 1992, but since then very little has been said or written in opposition to the

---

*Some of the most seriously endangered include the white-fin dolphin (whose population now numbers less than one hundred and is on the verge of extinction), Chinese sturgeon, Yangtze sturgeon, yanzhi fish, white dolphin, and river sturgeon.

"Wind box gorge." One of several smaller gorges inside Wu (witches) Gorge. At the center is the entrance to the "cavern of the enchantress." *(Photo by Audrey Topping)*

Table 1.1

Three Gorges Dam Specifications

Dam crest	185 m
Dam length	2,000 m
Reservoir Functions	
Normal pool level	175 m
Flood control level	145 m
Total storage capacity	39.3 billion m^3
Flood control storage	22.1 billion m^3
Navigation	Reservoir level raised by 10–100 m to allow 10,000–ton ships to Chongqing
Power Generation	
Installed capacity	17,680 MW
Unit capacity	26 units, 680 MW/unit
Inundation	
Land	632 km-long,19 cities, 326 towns
Arable land	430,000 mu [30,000 hectares]
Population	1,130,000 people

Note: Figures for land inundated and people moved are government estimates and are questioned by dam opponents.

Dam construction site near Sandouping (circa 1996). *(Photo by Richard Hayman)*

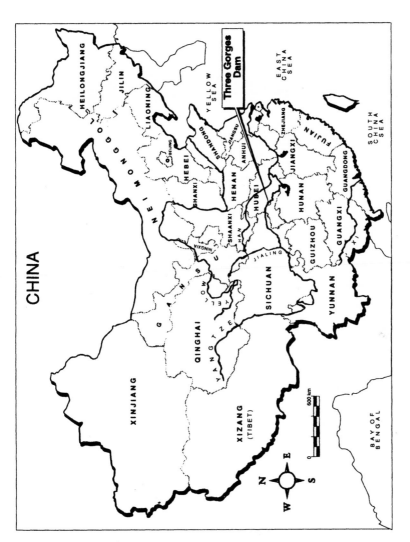

Map of China.

Pagoda near Wanxian; waters will flood to base of the structure. *(Photo by Audrey Topping)*

dam that will disrupt the lives of so many and damage such great swaths of our territory.*

Everyone knows that China is facing an energy shortage,** that our transportation systems are congested, and that we suffer frequent floods. The country has only recently emerged from the chaos of the Cultural Revolution (1966–76) and, with the pursuit since 1978 of a new, more open economic policy, increased foreign trade, and dramatic reforms in agriculture and commerce we have finally begun to experience some re-

---

* NPC approval of the dam came in April 1992, by a vote of 1,767 in favor, 177 opposed, and 644 abstentions, an unusual display of public opposition in the generally rubber-stamp body. The vote was in favor of a resolution to authorize construction of the dam and was conditioned on a promise from the Three Gorges Project Development Corporation to resubmit more precise construction schedules for future approval. See, Dai Qing, *Yangtze! Yangtze!*.

**In 1994, China generated 926 billion kilowatt-hours of electricity, 19 percent of which came from hydropower. Demand for electricity, which has been substantially underpriced in China's centrally planned economy, is expected to grow at an annual rate of at least 6 percent through the year 2000. See, *China Statistical Yearbook, 1996*, pp. 203–207.

markable economic gains. Why then, just when the country seemed to have a bit of money to spare, was this mammoth project proposed; especially when there were smaller and more viable options to meet our energy, transportation, and flood control needs?

The best alternatives involve building smaller dams on the Yangtze River's tributaries. But alternatives were never seriously considered by the top leadership. Why? Because China is in the midst of a phase of "uncontrolled" development where a sense of moderation and restraint are completely absent. This lack of control is evident at every level of planning for the Three Gorges project: From the "red specialists' " faith in technology, to the closed decision making of autocratic leaders, and the complete disregard for the environmental effects of the project on the river valley and its residents.

The Power of the Red Specialists

In China the so-called red specialists (*hongse zhuanjia*) consider themselves infallible even though the history of the People's Republic is littered with grandiose technological and economic projects gone wrong, often at enormous costs to the treasury and to human life.* With regard to the Three Gorges dam, this sense of infallibility manifests itself in a number of ways. For instance, the red specialists arrogantly claim that they have the technical ability and capacity to build the world's largest dam, turbines, and shiplift. But what they fail to consider is that the use of this technology does not make hydrological and environmental sense. Meeting the difficult technical challenges posed by the project should never take the place of sound scientific decision making. Decisions based only on what is technically possible may eventually succeed in building the dam and turbines, but they are unlikely to solve the pressing hydrological, environmental, and human problems which the dam will undoubtedly cause. This point was raised as early as the 1930s by Professor Huang Wanli. But unfortunately, the opinions of such venerable sages have had vitually no impact on policy that is driven by visions of technological grandiosity.

---

*Grandiose engineering and energy projects have also been criticized in the former Soviet Union. See, Grigori Medvedev, *No Breathing Room: The Aftermath of Chernobyl,* trans. Evelyn Rossiter (New York: Basic Books, 1993).

Even if the Three Gorges project is completed at the appointed hour, the long-term upheaval and damage caused by the resettlement of upward of 1.9 million people and the destruction of treasured cultural relics will be difficult, if not impossible, to reverse. The havoc created by the vast resettlement scheme will not only carry an immense price tag, but will also forever damage the spiritual and psychological health of the relocatees. The dam is not just about the loss of beautiful tourist landscapes, but about the damage the nation will do to itself through the patent disregard and ignorance of its spiritual wealth.

The "red specialists" have never managed to grasp the concepts of fundamental order and balance in the relationship between humankind and nature. At every turn—from its preference for a planned economy with a focus on iron and steel production, to its promotion of grain production, population growth, and large-scale dam construction*—the Chinese leadership has made decisions which run counter to the Chinese philosophical concepts of maintaining order and balance between humankind and nature. Not surprisingly, each of these decisions has caused immense damage to the country's environment and natural resources. For political reasons, however, those scholars and intellectuals who are in touch with this philosophical tradition have had very little opportunity to speak up. With the promotion of a new market economy since 1978, profit once again comes first in the minds of China's leaders, and all they think about is plundering nature rather than respecting and conserving it and maintaining the balance.

Adding to the problem is the fact that so many of these specialists make decisions based on blind self-interest, or on the narrow interests of their bureaucratic bailiwicks. A case involving the Leading Group for the Assessment of the Three Gorges Project is illustrative.** The youngest of the

---

*A reference to policies in the 1950s promoted by Mao Zedong over the objections of some scientists, agronomists, and hydrologists that led to converting almost all available land to grain production in order for each region to achieve agricultural self-reliance, that opposed population control on the grounds that more people meant more power for the "new China," and that led to a massive dam-building campaign during the Great Leap Forward (1958–60).

**Leading groups in China consist of a few to more than a dozen pertinent officials from various Party and government organs established to address a particular problem. The Leading Group for the Assessment of the Three Gorges Project was established in 1985 under the Ministry of Water Resources and Electric Power and led by Li Peng.

412 experts to advise the leading group was an unabashed supporter of the dam who longed to help with its construction. This young man was the only engineer out of the 412 researchers to be given the rank of specialist and was obviously promoted because of his unquestioning support for the project. By contrast, Huang Wanli, a hydrology and water resources expert and professor at Qinghua University, and Huang Shunxing, an agricultural and environmental protection expert, were barred from participating in the assessment of the project.

And then there's the case of Guo Laixi, one of nine experts who refused to sign their names to the assessment report. In a speech to one of the many meetings convened to assess the project, Guo noted that "China now confronts a very serious situation: There is a severe shortage of natural resources; our supplies of raw materials are seriously low; arable land is decreasing day-by-day; the population is growing rapidly; our agricultural base is extremely poor; the quality of the environment is deteriorating seriously;* inflation is widespread; financial deficits are growing; demand is outstripping supply; and the economic reforms are confronting many difficulties. Any proposal for an early and speedy launch of the project under these conditions is," Guo continued, "not a simple error or an example of negligence, but a calculated preference on the part of the leadership. For if the Three Gorges project is not launched immediately, the authorities will have to find jobs for the tens of thousands of workers who will soon be out of work when the Gezhouba dam is completed.** In other words, the decision to launch the project is really about meeting the personal interests of workers and their families in the

---

The leading group oversaw and coordinated the fourteen studies that comprised the assessment report for the Three Gorges dam, which preceded a full decision by the NPC. See, Kenneth Lieberthal, *Governing China: From Revolution Through Reform* (New York: W.W. Norton & Co., Inc., 1995), p. 193., and Dai Qing, *Yangtze! Yangtze!*, p. 18.

*China's environmental problems are analysed in Vaclav Smil, *China's Environmental Crisis: An Inquiry into the Limits of National Development* (Armonk, N.Y.: M.E. Sharpe, Inc., 1993), and He Bochuan, *China on the Edge: The Crisis of Ecology and Development*, ed. Xu Yinong (San Francisco: China Books and Periodicals, Inc., 1991).

**Located 40 kilometres downstream from the current site of the Three Gorges dam at Sandouping, the Gezhouba dam was supposed to take five years to build and cost ¥1.35 billion. Instead, the project took nineteen years to build and ultimately cost ¥5 billion (¥:$ = 8:1).

Gezhouba dam. *(Photo courtesy of Jim Williams)*

various construction gangs and organizations that make up the dam-building industry in China."

Reckless actions by specialists and bureaucrats who possess narrow scientific and technical skills can be very frightening indeed. Such people plan things in very meticulous ways to fit their own personal interests and work only to advance the goals of their respective bailiwicks. They could care less about the national interest and the fate of the nation.

Autocracy and Closed-Door Decision Making

Throughout its history, China has been ruled by an autocratic system. In the distant past, everything was done in the name of the emperor. After the 1912 Republican revolution, it was done in the name of the people's revolution. And since 1949, during the reigns of Mao Zedong and Deng Xiaoping, it has been done in the name of communism and socialism. Autocracy is still considered an acceptable form of government in some areas and under some circumstances either because there is no alternative system, or because it is believed to be appropriate at a certain stage of a nation's development. Nevertheless, autocratic governments are on the wane.

The Three Gorges project has both benefited from China's autocratic history and helped strengthen it. Those promoting the dam, from the 1950s to today, have all been masters of political gamesmanship, constantly referring to "Chairman Mao's desire" (*Mao zhuxide xinyuan*) and "Deng Xiaoping's support and concern" (*Deng Xiaopingde zhichi he guanxin*) for the project. By invoking the support of the country's autocratic leaders, the dam was made virtually unassailable.*

When the project did run into resistance, the dam-supporters used nationalistic bluster to reinforce their position. Nationalism is an inherently parochial, irrational, and extremely destructive force that ultimately runs counter to the interests of human development. It should only be called on in extreme circumstances, such as in resisting foreign invasion, and not otherwise used to stir passions and excitement.

Although private companies and other ostensibly private organizations have been established to assist in the construction of the dam, the project has relied on government financing since its inception. Given that China is trying to move in the direction of a market economy, the decision to build a large project such as the Three Gorges dam solely on the basis of the leadership's will can only have a negative impact on the transition.

Government munificence has come in many forms: direct allocations by the state; the transfer of revenues from the Gezhouba dam; and increases in national electricity rates. The government has also "recommended" that some profitable large enterprises "assist their counterparts" through donations to the Three Gorges project. This sort of action strengthens and supports the central planning apparatus in the economy and works to stifle independent thought and competition.

Because local leaders are centrally appointed under China's autocratic system, they do not dare strive for a fair deal for their local constituencies. The people of Chongqing, Sichuan (who will receive few if any benefits from the dam and may suffer many of its negative effects), have con-

---

*In 1953, Mao first expressed interest in the Three Gorges dam and insisted on building a single large dam, instead of a series of smaller ones on the Yangtze's tributaries, something that had been proposed by the hydrologist Lin Yishan. Mao even suggested that he might resign as chairman of the Chinese Communist Party to assist in the project design which was eventually overseen by Zhou Enlai. *Mao Zedong zai Hubei* (Mao Zedong in Hubei Province) (Wuhan: Hubei People's Publishing House, 1993), pp. 95–100.

demned their leaders for selling out Sichuan's interests.* Even more significant is that, in 1989, amid strong opposition to the dam, the State Council decided to postpone consideration of the project. But in the political atmosphere following the Tiananmen Square massacre, all opposition to the project in the government was crushed, and "senior leading cadres" used their political weight in the traditional style of autocratic politics to ignore legal procedures and ensure that the project went forward.[1]

Subsequently, when the Three Gorges project was awaiting approval from the NPC, the national press was mobilized to write only positive reports about it. Meanwhile, even before the NPC convened for its vote, the chair made it clear that its approval was not in question.[2] During the course of the session itself, the microphones on the floor of the NPC were turned off to prevent the dam-opponents among the delegates from voicing their views and generating collective opposition.**

China's autocratic leaders have used the most undemocratic procedures imaginable to push the project forward. I don't think for a moment that China's modernization can be achieved overnight, but the government and the people should break with the traditional autocratic system and make a conscious effort to gradually begin the transition to a more open system in order to bring about a fundamental transformation in China's political culture. Instead, supporters of the Three Gorges project continue their efforts to consolidate power and support the old system by whatever means necessary in order to ensure that the construction goes forward.

The Effects of Uncontrolled Development on the Environment

Even if construction of the Three Gorges dam is completed as planned in 2013, its ability to generate electricity depends on avoiding a massive

---

* In 1996, Chongqing Municipality was granted province-level status under the direct authority of the central government, thereby separating it from Sichuan Province, whose leaders have generally not supported the dam. Province-level conflicts and divergent interests over the dam are analysed in Kenneth Lieberthal and Michel Oksenberg, *Policy Making in China: Leaders, Structures, and Processes* (Princeton: Princeton University Press, 1988). See also, Epilogue.

**These events are described in Dai Qing, *Yangtze! Yangtze!*, pp. 107–117.

Table 1.2

Construction Phases

Phase	Year	Construction stage	Water level
Preparation 1993			
First 1994–97	1994	Excavation of base begins Project inaugurated	
	1995	Pouring of reinforced concrete begins	
	1997	River blocked and diverted	
Second 1998–2003	2003	Electricity generation begins	135m
Third 2004–2013	2007		156m
	2009	Electricity generating system completed	
	2013	Normal operation	175m

buildup of sediment behind the dam. Because of sedimentation, the Three Gate Gorge dam (*Sanmenxia*) on the Yellow River has induced floods in the river's upper reaches and led to the resettlement of over 400,000 people. It now produces less than one-third of the power that was promised, its turbines are damaged by sediment, and it will not be able to fulfill its flood-control function until another massive dam, the Xiaolangdi, is built downriver.*

The Three Gorges dam will face similar sediment-related problems. Even if the dam does generate the promised electricity, most of it will go to serve southern and eastern China. Sichuan Province will be unable to develop its own regional electrical supply because all of the money available for electricity generation is tied up in the Three Gorges project. The province will reap few benefits from the dam, but will bear many of its costs, especially the loss of land and the burden of resettlement.

The primary purpose of the Three Gorges dam is flood control, and it

---

* After the Three Gorges dam, the Xiaolangdi dam is the second largest such project in China. Slated for completion in 2002, it will cost U.S.$3 billion and will involve the relocation of over 400,000 farmers.

has been designed to contain a once-in-one-thousand-year flood. But no single dam could ever contain such a flood on the Yangtze River. Unable to contain massive floods, the Three Gorges dam provides, conversely, an excessive and unnecessary level of protection from the smaller floods which frequent the Yangtze. Even at its peak, the 1981 flood in Sichuan Province never reached the cities of Yichang or Wuhan.*

From the beginning of the dam project, Huang Wanli has consistently warned the leadership against creating a situation similar to the "Railroad Protection Movement in Sichuan" which, he noted, "led to the 1912 Republican revolution."** That revolution, we now know, turned out to be enormously destructive. People in China and throughout the world sincerely hope that the country's transformation and modernization can be carried out smoothly, but the Three Gorges project runs counter to this hope because, in its name, the government has suppressed free speech and strengthened its power at the expense of the provinces and the people. The project is encouraging corrupt economic practices in enterprises and in the government and will lead to an enormous waste of resources, all the while destroying the environment and violating the rights of the people.

We are fortunate that we live in an open world, for the effects of the Three Gorges project transcend national boundaries. If the project is to be supported financially by multinational organizations, then it cannot avoid the scrutiny of the outside world.

The human race has readily demonstrated its capacity to destroy the environment, and we do not yet know how to control our desires and greed. So what should we do when such an uncontrolled project is being carried out under the watchful eye of the Chinese public? I know that other countries subject their hydropower projects to public scrutiny with success. But how can the Chinese people struggle for the same assurances in the case of the disastrous Three Gorges dam?

---

*Evidence also exists that Chinese government officials have grossly exaggerated the severity of recent floods to justify construction of the Three Gorges dam. See, Simon Winchester, *The River at the Center of the World: A Journey Up the Yangtze, and Back in Chinese Time* (New York: Henry Holt and Company, 1996), pp. 220–230.

**The Railroad Protection Movement was an immediate cause of the 1911 revolution that overthrew the Qing (Manchu) dynasty (1644–1911). The movement was centered in Sichuan where local merchants resisted the central government's railroad nationalization plan because it entailed foreign loans, fostered official corruption, and led to the imposition of commercial taxes to finance the entire scheme.

Notes

1. This information was contained in a summary of the Lunar New Year Forum published in *Xinhua Monthly,* which covers domestic developments. [Note: Many sources provided in the original text are incomplete, eds.]

2. According to a participant at the meeting who wishes to remain anonymous.

Chapter Two

A Profile of Dams in China

Shui Fu

The Great and Small Leaps Forward

There were virtually no large-scale water projects in China before 1949. But in the ensuing years, and especially in the years during and since the Great Leap Forward (1958–60), the Chinese Communist Party has heavily promoted dam and reservoir construction as part of massive national campaigns. In less than forty years all of China's major rivers have been dammed. In the mind-set of the Chinese people, dam projects became more than just another kind of construction project; the campaigns promoting dam construction equated harnessing rivers with developing the country and mandated absolutely that citizens demonstrate their "positive support" and "political enthusiasm" for the projects. Under the dictatorship of the Party, the goal of "harnessing water" became equated with "harnessing people."

Most people associate the "Great Leap Forward" with national campaigns to increase iron and steel production. But just as important was "engaging the mass movement" for the "large-scale water conservancy campaign." At that time, water conservancy policy "gave primacy to the accumulation of water and to irrigation and gave only secondary consideration to drainage and flood control" (*yixu weizhu, xuxie jianchou*).* There was, however, lively debate on the subject. One school favored relying on local initiatives to build small-scale dam projects which would emphasize the accumulation of water for irrigation purposes. The other favored state-sponsored, large-scale projects whose primary function would be flood control.

---

*During the Great Leap Forward, Chinese leaders thought that by 1972 hydropower would produce more than half of the country's power. See Michel Oksenberg, "Policy Formulation in China: The Case of the 1957–58 Water Conservancy Campaign" (Ph.D. diss., Columbia University, 1969), and Lieberthal and Oksenberg, *Policy-Making in China*, p. 96.

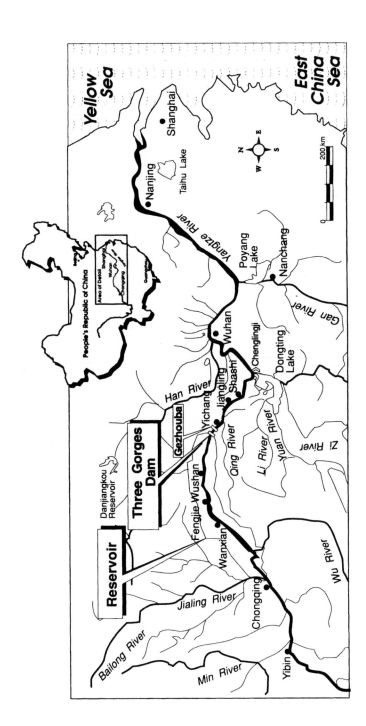

The Yangtze River and all its major tributaries.

The debate lasted for decades and was transformed from an academic dispute over the merits of the different approaches into a political struggle in which the supporters of the first approach won a decisive victory. As a result, "accumulation" (*xu*) was taken to the extreme. Anhui Province built an excessively large "river irrigation network," while people in the north took the policy of accumulating rain water for irrigation to such extremes that their fields became waterlogged. Thus the "water conservancy campaign" was ultimately reduced to a campaign to build reservoirs and dams, and by 1990, 83,387 of them had been built in China.* Three hundred and sixty six of them had a capacity over 100 million cubic meters, 2,499 had a capacity of 10 million to 100 million cubic meters, and more than 80,000 had capacities below 10 million cubic meters.**

In the early 1970s, during the Cultural Revolution, there was a second, smaller "Leap Forward" of water conservancy projects in which dam and other irrigation projects once again began sprouting up all over the country.

The first dam construction boom in the 1950s was a thrilling time. People's communes, the Great Leap Forward, and the manufacture of iron and steel all stimulated the construction of more hydropower projects. Leaders boldly approved projects to accumulate more water for irrigation without knowing whether they were feasible. According to He Xiaoqiu, the former deputy chief engineer of the Hydropower Investigation and Design Institute of the Ministry of Water Resources and Electric Power,† all a particular leader had to do was point his finger at a certain place and

---

*This policy represented a triumph for a radical view of water conservancy policy that downgraded the role of scientists and technicians and expanded the power of local Communist Party officials over decisions involving dam and reservoir construction. It replaced a more moderate policy which, in 1956–57, had emphasized repairing existing facilities and giving primacy to drainage and soil conservation while also granting authority over water conservancy management to technicians. See, Oksenberg, "Policy Formulation."

**China Water Conservancy Yearbook* (Zhongguo shuili nianjian) (n.p., n.d.).

†The Ministry of Water Resources and Electric Power has been a strong supporter of the Three Gorges dam, overseeing both the studies of the Leading Group for the Assessment of the Three Gorges Project and the work of the Yangtze Valley Planning Office, which is responsible for the overall plan for developing the Yangtze River. From 1979 to 1982, and again after 1988, the Ministry of Water Resources was split from the electric power division, which currently resides in the Ministry of Power. See Lieberthal and Oksenberg, *Policy Making in China,* pp. 94–102, 283–287.

the decision would be made to build a dam between one mountain and another. The engineers were left to assess whether the project made sense, but few projects were rejected: No one wanted to be accused of being a "rightist" or "an obstruction on the bridge leading to communism."*

As a result, reservoirs for irrigation were built *en masse.* "There was water as far as the eye could see," said He Xiaoqiu on visiting one of the country's largest reservoirs, which, to his horror, was being operated by a teenage girl who had just graduated from the hydro training program. The government's slogan "The land will yield as much grain as the people desire" was being taken to heart—projects were being built (and operated) with abandon.

But some were concerned about the emphasis on accumulation and irrigation over all other goals, and especially about its effects on the landscape. After the 7,000 Cadres Conference in 1962,** Zhou Enlai openly expressed his concern about the situation. "I've been told by doctors that if a person goes without eating for a few days, no major harm will result. But if one goes without urinating for even one day, they will be poisoned. It's the same with land. How can we accumulate water and not discharge it?" In 1964, Zhou pushed for a more comprehensive approach. He proposed a management policy for water conservancy projects that was very similar to the existing policy, but with one important exception: He recommended that all aspects of water conservancy projects be integrated and that the projects be governed by a comprehensive approach to management. Senior hydrologists correctly understood Zhou's policy as an attempt to make the previous policy somewhat ambiguous and to focus on the comprehensive management of water conservancy projects, whatever their size. In 1966, Zhou also commented: "I fear that we have made a mistake in harnessing and accumulating water and cutting down so much forest cover to make way for more agricultural cultivation. Some mistakes can be remedied in a day or a year, but mistakes in the fields of water conservancy and forestry cannot be reversed for years."

---

*The 1957 Anti-Rightist Campaign, which preceded the Great Leap Forward, targeted scientists and intellectuals for persecution.

**Held in January and February 1962, this conference convened in the aftermath of the disaster created by the Great Leap Forward and noted that people who had spoken the truth about the Leap were punished and people who submitted false reports and hid defects were rewarded. See Kenneth G. Lieberthal and Bruce J. Dickson, *A Research Guide to Central Party and Government Meetings in China: 1949–86* (Armonk, N.Y.: M.E. Sharpe, 1989), p. 126.

However, no one, not even Zhou Enlai, was able to block the [dam-building] "campaign." In 1958, hydro departments in the Ministry of Water Resources and Electric Power established the Office of the Water Conservancy Campaign (*Shuili yundong bangongshi*). According to Liu Derun, the then deputy director of this new office: "Our daily work consisted of making phone calls to the provinces inquiring about the number of projects they were building, how many people were involved, and how much earth they had moved. In hindsight, some of the data and figures we gathered were obvious exaggerations, but no one back then had the energy to check them out." From 1949 to 1959, 800 million cubic meters of earth was moved—580 million cubic meters in 1958 alone.

Are Water Conservancy Projects Beneficial or Harmful?

Before 1949, only twenty-three large- and medium-scale dams and reservoirs existed in all of China. One of the earliest was built in Sichuan by Li Bing during the Qin dynasty (221–207 B.C.). Another, the Fushan dam on the Huai River, was used to block passage across the Huai during an attack against the Wei Kingdom in A.D. 516. The scale and sophistication of the Fushan dam were unprecedented for the time, but the knowledge gained through its construction was not passed down: Like so much else in China, it disappeared with the collapse of the imperial autocracy. The Fushan dam also demonstrated to the world the kind of disasters that large dams can produce.*

The more than 80,000 dams and reservoirs built over the last forty years have played an important role in flood control, electricity generation, and irrigation and have provided water for urban areas and industry. These achievements should not be underestimated, but dam construction, especially during and after the Great Leap Forward, has also had disastrous consequences.

By 1973, 40 percent or 4,501 of the 10,000 Chinese reservoirs with capacities between 10,000 and one million cubic meters were found to have been built below project specifications and were unable to control floods effectively. Even more dams had problems relating to the geology of the dam site, and to sedimentation. Most serious, however, were the

---

*Four months after the dam's completion, the Huai overtopped the Fushan releasing 10,000 million cubic meters of water and killing 10,000 people downstream. See, Nicholas J. Schneider, *A History of Dams: The Useful Pyramids* (Rotterdam: AA Balkema Publishers, 1994), pp. 102–103.

numerous dam collapses. By 1980, 2,976 dams had collapsed, including two large-scale dams [the Shimantan and Banqiao dams]. One hundred and seventeen medium-sized, and 2,857 small dams had also collapsed. On average, China witnessed 110 collapses per year, with the worst year being 1973, when 554 dams collapsed. The official death toll (not including the Banqiao and Shimantan collapses*) resulting from dam failures came to 9,937. Some people say that among the more than 2,000 dam collapses, only 181 involved fatalities but this hardly seems accurate.

By 1981, the number of formally recognized dam collapses had risen to 3,200, or roughly 3.7 percent of all dams. According to Ma Shoulong, the chief engineer of the Water Resources Bureau of Henan Province, "The crap from that era [the Great Leap Forward] has not yet been cleaned up." In 1958, more than 110 dams were built in Henan; by 1966 half of them had collapsed. Of four key dams on the Yellow River—the Huayuankou, Wei Mountain, Luokou, and Wangwang Village dams—two were dismantled and two were postponed.

Many of the dams which remain are unsafe and in need of repair. A 1985 study claimed that one quarter of all dams fell into this category, and by 1986 the government had singled out 43 especially dangerous dams because they threatened major towns, industries and mines, major transportation routes, and military facilities. Thirty-five of these so-called "Category 1" dams were large-scale, and eight were medium-sized. By the end of 1992, 30 had been repaired or reinforced. In 1992, a "Category 2" ranking was established for 38 dams which were to be reinforced during the "Eighth Five-Year Plan" (1991–95) or later.

According to experts, if the riskiest of these dams were to fail, hundreds of thousands of people could be killed. But current levels of funding are woefully inadequate to repair or reinforce the dams. At least ¥5 billion would be required for the large- and medium-sized reservoirs alone. Where will the money come from? According to Vice Premier Tian Jiyun,

---

*According to the Ministry of Water Resources and Electric Power, over 20,000 people lost their lives as a result of the Banqiao and Shimantan collapses. However, the National Storm Flood Investigation Group of the Ministry of Electric Power and the Water Data Research Group of Nanjing put the figure at 85,000 in their book titled *Zhongguo lidai dazhi shui* (Harnessing Water Throughout the Chinese Dynasties) (n.p., 1988). Eight key dam critics and members of the Chinese People's Political Consultative Conference (a largely ceremonial body that meets prior to the convening of the NPC) estimated that the true death toll was 230,000. For more on the dam collapses, see Chapter Three.

the task must be completed according to schedule or those in charge will be held accountable. The Ministry of Water Resources, however, just shrugs its shoulders. Everyone knows the task is impossible. It would appear that the "crap" left by the Great and Small Leaps Forward will linger for some time to come.

It is difficult to predict the disasters that these dams might produce should they fail, because most information regarding dam collapses in China is confidential. During a 1991 conference on dam collapses in Vienna, participating countries exchanged information, as is the general practice, on collapses in their respective countries. Only China indicated that it had no dam collapses to report. Foreign experts attending the conference commented to China's representative, Pan Jiazheng, that it was miraculous for a country as big as China, a country with more than 80,000 reservoirs, to have had no dam collapses. Either our representative knew nothing about the dam collapses or, owing to Party discipline, he could not say. All in all, he must have been very embarrassed.*

---

*The meeting mentioned here would appear to be the 1991 meeting of ICOLD, the International Conference on Large Dams.

Chapter Three

The World's Most Catastrophic Dam Failures

The August 1975 Collapse of the Banqiao and Shimantan Dams

Yi Si

In August 1975, a typhoon struck Zhumadian Prefecture of Henan Province in central China, causing reservoirs to swell with rainwater behind dozens of dams.* When the torrential rains subsided, the massive Banqiao and Shimantan dams had collapsed, as had dozens of smaller dams. The destruction downstream was unprecedented: Eighty-five thousand people were dead, and millions more lost their homes and livelihoods.

The storm began on August 4, when the typhoon skirted by Taiwan and hit the Chinese mainland at the Jin River in Fujian Province. Because of a meeting of unusual weather patterns—one originating in the southern hemisphere near Australia and the other from the Western Pacific—the typhoon did not expend itself quickly when it reached Fujian, as is usually the case with storms coming in from the South China Sea. Instead, it gathered force as it moved through the southern provinces of Jiangxi and Hunan, and then took a sharp northerly turn straight toward the Yangtze River and the central plains.

On August 5, the typhoon suddenly dropped off the radar screen at the

---

*Excerpts from this report appear in Human Rights Watch/Asia, "The Banqiao and Shimantan Disasters." The typhoon's official designation was "Number 7503," and the events it helped precipitate, were called the "August 1975 Disaster."

Central Meteorological Observatory in Beijing. It had shifted direction again and was moving east, where it crossed over Henan Province heading toward the Banqiao and Shimantan dams.

The storm hit hardest in the valleys between the Funiu and Tongbai mountains, where eyewitnesses described rainfall that had the force of a fireman's hose, and where, after the rain had subsided, dead birds were found strewn everywhere, the hapless victims of raindrops falling with the force of arrows. With some irony, the valleys, because they were prone to heavy rainfall, had always been considered an ideal place to build reservoirs. By 1975, when the typhoon hit, there were more than 100 of them in the area. But the dams could not withstand a storm this severe—one which dumped 1,000 millimeters of rain in just three days and which was well beyond the worst-case scenarios imagined by the designing engineers.

On August 8, the Banqiao reservoir at the Ru River was at maximum capacity; the water had reached the crest of the 118–meter dam and was still rising. The dam could not release water as fast as its reservoir was filling. When, shortly after 1:00 A.M., the flood waters rose 30 centimeters above the crest, the main part of the dam gave way and 600 million cubic meters of water surged forth. A wall of water six meters high and 12 kilometers wide rushed into the river channel and surrounding valleys and plains and destroyed virtually everything in its path.

The smaller of the two reservoirs, the Shimantan, reached its maximum capacity a half hour before the Banqiao, at 12:30 A.M. When the water rose 40 centimeters above its crest, the Shimantan dam also collapsed. One hundred and twenty million cubic meters of water burst forth from the dam at a rate of 25,300 cubic meters per second. Within five hours the entire reservoir was virtually empty.

The Nihewa and Laowangpo flood diversion areas downstream of the dams could not handle the 720 million cubic meters of water rushing out of the reservoirs; their capacity was only 426 million cubic meters. In Zhumadian Prefecture, many of the dikes collapsed, creating a 300– by 150–kilometer lake. When the Nihewa and Laowangpo catchment areas gave way on August 8, 100 million cubic meters of water poured into the Fenquan River. Later, on the evening of August 9, the floods reached the Fuyang area in Anhui Province. The Quan River dikes collapsed and the entire Linquan county seat was submerged.

According to the former minister of water resources and electric power, Madame Qian Zhengying, the disaster flooded 29 counties and

The quest to harness rivers is not a recent one in China. During the Former and Later Han dynasties (202 B.C.–A.D. 8 and A.D. 25–220, respectively) irrigation projects proliferated. The Qingpo and Gepo projects in Xincai County, the Hongxibo and Shitangpo projects in Runan County, and the Ju Marenpo project in Biyang County are famous irrigation projects from this period. Numerous dams were also built from the Three Kingdoms (A.D. 220–280) to the Sui (A.D. 581–618) and Tang (A.D. 618–906) dynasties. These included the Ershisi in Xiping, the Zhangze in Suiping, the Shen in Zhengyang, and the Huangling in Shangcai. In the thousand years or so since the Song dynasty (A.D. 960–1279), however, most of these water control projects fell into disrepair and deteriorated badly, often contributing to floods rather than controlling them.

During the thirty-year campaign to "Harness the Huai River," the Shimantan dam was built in the upper reaches of the Hong River, and the Banqiao dam was built in the upper reaches of the Ru River. At that time, very little was known about the local hydrology and, as a result, there were serious problems with the original design and construction of the dams. During the second stage of construction of the Banqiao, cracks were discovered in the sluice gates and in the dam structure itself. As a result, in 1955–56, the two dams were reinforced and expanded using Soviet expertise.* The Banqiao dam was designed to protect downstream areas against severe floods expected only once every 100 years, and its spillway was designed to pass floods expected once every 1,000 years. A 1,000–year flood would cause 330 million cubic meters of runoff from 530 millimeters of rain over three days, and would create a peak inflow to the reservoir of 5,083 cubic meters of water per second. Based on these criteria, it was decided that the dam height would be increased by three meters to 116.34 meters and that an auxiliary spillway would be built. So, along with the original water release channels and sluice gates, the maximum flood spillway discharge capacity was set at 1,742 cubic meters of water per second with a maximum storage capacity for the reservoir of 492 million cubic meters and a flood storage capacity of 375 million cubic meters.

The Shimantan dam was designed to protect downstream areas against

---

*For a study of Soviet dam design in the 1930s, see Anne D. Rossweiler, *The Generation of Power: The History of Dneprostroi* (Oxford: Oxford University Press, 1988). Such problems led to an increasing concern in the mid-1950s among Chinese hydrologists about the quality of Soviet advice.

a once-in-50–year flood, with spillways designed for a once in 500–year flood. It would cause 88 million cubic meters of runoff from 480 millimeters of rainfall over a three-day period, and create a peak inflow to the reservoir of 1,675 cubic meters of water per second. Based on that design, the dam height was increased by 3.5 meters to 109.7 meters, the maximum storage capacity was 94.4 million cubic meters with a flood storage capacity of 70.4 million cubic meters.

After the Banqiao and Shimantan dams were built, the pace of dam construction quickened and was extended beyond the mountain regions and into the central plains. Between 1957 and 1959, over 100 dams and reservoirs were built in Zhumadian Prefecture alone. Henan Province's experiences in the "Harness the Huai River" campaign became the model for dam construction nationwide during the Great Leap Forward in 1958. Vice Premier Tan Zhenlin announced a policy of "giving primacy to water accumulation for irrigation, construction of small-scale reservoirs, and economic self-reliance of the agricultural brigades." Since the problems of dam and reservoir construction in the mountains "had basically been solved," he proclaimed that "dam building should be extended to the plains."

In response to the vice premier's plans, a hydrologist named Chen Xing pointed out that building reservoirs in the plains and the policy of giving "primacy to accumulation and irrigation" would cause serious environmental damage. He argued that the accumulation of vast quantities of water in reservoirs throughout Henan Province would raise the water table beyond safe levels, cause severe waterlogging of agricultural land, and increase the salinity and alkalinization of the soil. In short, he warned that dam construction on the plains would have disastrous consequences.

Unfortunately, these warnings were ignored. The authorities' fixation on the policy of giving primacy to the accumulation of water and to irrigation came to define dam-construction policy in China and led to the construction of reservoirs and dams on a massive scale. In Anhui Province a whole slew of small-scale reservoirs soon sprang up in hilly areas. And the Huai River Valley was dammed to the point where the river's natural flow was essentially destroyed.

Chen Xing designed the largest reservoir on the plains, the Suya Lake reservoir of Zhumadian, Henan, which was built during the Great Leap Forward. During its construction, a deputy head of the Henan Province water resources department criticized Chen's design as being "too conservative" and, in defiance of hydrological safety standards, reduced the number of sluice gates in the dam from an originally planned twelve to only five.

Similarly, officials reduced the number of spillway gates at the Bantai emergency flood diversion project on the border of Henan and Anhui provinces from the original nine to seven. Then, in 1961, they blocked off two of the remaining gates, all under the guise of "primacy to accumulation and irrigation."

Following the Great Leap Forward disaster, Henan, and the country as a whole, experienced the "three difficult years" (1959–61).* According to accounts at the time in the Chinese press, this was supposedly not a difficult time for Henan, and no large-scale disasters were reported in the area. But in reality, flooding and famine were widespread throughout the central plains area. Liu Jianxun, who was appointed Party secretary in Henan Province in 1961, blamed dam and reservoir construction for the disasters. To help correct the problems, he searched everywhere for the dam "opponents" and found Chen Xing at Xinyang.

During the Great Leap Forward's backyard furnace campaign, Chen had suggested that donating watches and other such personal items to the steel campaign was useless and that the price of the campaign was too high. He also challenged the idea that the "Sputnik model" of the People's Commune (with its extravagant claims of vast increases in agricultural production) should be emulated by an equally extravagant policy of building more and bigger dams. For this he was labeled a "right-wing opportunist," and purged.

During the 1961 campaign to "rectify deviations and correct past errors," a large-scale study of dams and reservoirs was conducted in Henan. A number of substandard dams were repaired and many other potentially dangerous ones, including a few large dams, were demolished. The policy of "primacy to water accumulation and irrigation" was not, however, completely reversed. By the end of the 1960s, the persistence of this policy led to the construction of more than 100 additional dams and reservoirs in Zhumadian. The reclamation and settlement of large tracts of land which had historically been left bare for flood diversionary purposes further

---

*The disastrous impact of the Great Leap Forward on Henan Province is documented in Jean-Luc Domenach, *The Origins of the Great Leap Forward: The Case of One Chinese Province*, trans. A.M. Berrett (Boulder: Westview Press, 1995), and in Jasper Becker, *Hungry Ghosts: Mao's Secret Famine* (New York: Free Press, 1997).

reduced the ability to manage emergency floods on the Hong and Ru rivers. Other flood management measures, such as the strengthening of dikes, were also neglected because no one even considered the possibility of a catastrophic dam collapse. Just prior to the 1975 disaster, a 1.9–meter-high earthen embankment was added on to the top of the Shimantan dam to increase its overall holding capacity, while at Banqiao, officials authorized an additional retention of no less than 32 million cubic meters of water in excess of the dam's designed safe capacity.

The dam's modifications and the persistence of the policy of accumulating water above all else were based on limited historical knowledge and incomplete survey data.* When the big flood hit in August 1975, they were dumbfounded by the torrential rains.

"The River Dragon Has Come (*Chu Jiaozi!*)"

The "August 1975 Disaster," as the dam collapses have come to be known, was the result of three successive deluges. The storm hit southern Henan on August 5, and in the initial downpour, a total of 448.1 millimeters of rain fell on the region, about 40 percent more than the heaviest previous rainfall on record. The water level at the Banqiao dam rose to 107.9 meters, bringing it close to maximum capacity.

The experts were now at the mercy of the flood: The sluice gates were opened, but were partially blocked by accumulated silt. This impeded the dam's ability to drain, so the water level continued to rise causing flooding in nearby areas. By the evening of August 5 the water was already one meter deep in the courtyard of the Banqiao Reservoir Management Bureau. The flooding knocked out telephone service causing the bureau to lose contact with the weather stations measuring rainfall in the upper reaches of the reservoir. As the waters continued to rise, all telephone and road communication with the dam site was severed. The cadres of the Banqiao commune tried to move the elderly and the children out of the area and to rescue what files they could.

The second deluge began at noon the following day and lasted for sixteen hours. The water at the dam reached 112.91 meters, more than two meters above its designed safe capacity. The third and final downpour

---

*The woefully inadequate data on local conditions is documented in Oksenberg, "Policy Formulation."

began at 4:00 P.M. on August 7 and lasted for thirteen hours. That evening at 7:00, the Zhumadian Municipal Revolutionary Committee convened to assess the dangers posed by flooding to the dams at Suya Lake, Songjiachang, Boshan, and elsewhere. The potential danger to the Banqiao dam, however, was never mentioned: it was thought to be an "iron dam," one that could never collapse. By 9:00 P.M., seven smaller dams at Queshan, Miyang, and elsewhere in the area had collapsed, followed an hour later by the medium-sized Zhugou dam. Before the sun rose the next morning, 62 dams had collapsed in Henan.

Around the same time, a thin line of people stood strung out across the top of the Banqiao dam, toiling waist-deep in water trying to repair the rapidly disintegrating embankment. Suddenly there was a flash of lightning followed by deafening thunder. Silence followed, and, for a brief instant, the skies cleared and the stars appeared again overhead. Someone shouted, "The water level is going down! The flood is retreating!" But just a few seconds later, the dam gave way, and 600 million cubic meters of water erupted with a demonic and terrifying force. Somewhere, a hoarse old voice cried out, "The river dragon has come!"

Over the next five hours, a gigantic wall of water travelling nearly fifty kilometers per hour cascaded downward over the surrounding valleys and plains obliterating everything in its path. Shortly afterward, the Shimantan dam also collapsed, to largely similar effect. Entire villages and small towns disappeared in an instant. A government order issued the previous day to evacuate local residents had applied only to populations living in the immediate vicinity of the Banqiao dam; eastward of Shahedian Town, no such evacuations had been carried out. Shahedian Town saw 827 of its 6,000 residents perish. Wencheng commune, east of Shahe County, lost half of its 36,000 residents, while 1,000 of the 1,700 residents of the commune's Weiwan Brigade were killed. Wei Shixing, a peasant who lost two family members in the storm, recalled, "I was looking for rope to help get my parents on to the roof of our house. But as soon as my father tied the rope to the roof, the flood waters rushed into our house and snapped the thick rope like a thread. The last thing I saw was my parents being carried away by the water."* The massive Suya Lake Reservoir, whose

---

*See, *Zhongguo lishi dahongshui, xiace* (Major Floods in Chinese History, Vol II), p. 141. This book contains the only known photographs, taken on the day after the dam collapses, of the post-disaster scenes at Banqiao and Shimantan. A newsreel of the disaster was also produced for viewing by China's top leadership, but, as far as is

emergency sluice gates had been more than halved in number by ardent Maoist officials many years earlier, successfully withstood the typhoon, but thanks only to remedial construction work that had been completed a mere eight days earlier.

At 8:00 P.M. on August 9, Zhumadian Prefecture sent an emergency telegram to the central government that read as follows:

> *To the Central Committee of the CCP, the State Council, and the Central Military Commission*: From August 5 to 8, there were a number of severe rainstorms in our area with an average rainfall of 800 millimeters that caused flood waters to accumulate to a depth of two meters like an ocean. The Banqiao Reservoir collapsed at 1:00 A.M. on August 8. The Suiping county seat was submerged and many people died. As a result of the storm, three million people have been engulfed by the flood and have been isolated on the tops of roofs and in trees for three days. Emergency!

The aftereffects of the disaster were, if anything, worse than the disaster itself. The water released from the collapsed dams combined with entrapped localized flood waters to form a huge lake, thousands of square kilometers in size, which submerged or partially covered countless villages and small towns. Because of the decades-old neglect of dike maintenance, flood diversion systems, and dredging programs in the region, there was nowhere for the water to go. The complete disruption of all transportation and communications in the region also meant that the army's emergency crews were unable to reach, feed, clothe, or otherwise assist most of the survivors for up to two weeks after the initial disaster. People were stranded in the water for days, having to endure scorching sun and blistering heat. The medical teams were helpless in the face of the catastrophic health epidemics that swiftly ensued. I learned about the aftereffects of the dam collapses by reading through documents and files from Zhumadian Prefecture, which included the following telephone records:

> August 13: East of Xincai and Pingyu, the water is still rising at a rate of two centimeters an hour. Two million people across the district are trapped by the water. . . . In Runan, 100,000 people who were initially submerged

---

known, it has never been viewed by anyone outside the country. Coverage of the disaster in the Chinese press at the time was limited to stories of "heroic" efforts by rescue teams and locals residents, with little information on the extent of the disaster or loss of life. See *Henan ribao* (Henan Daily), August 26, 1975, and September 14, 1975. A similar coverup of the Chernobyl catastrophe is noted in Medvedev, *No Breathing Room*.

but somehow survived [by clinging to trees, rooftops, etc.] are still in the water. Forty thousand people have been rescued; 200,000 are sick with diarrhea and other related illnesses. There's no medicine. In Shangcai, 300,000 people are marooned on the dam, on rooftops, and elsewhere. Twenty communes have been engulfed by floodwaters. Many people haven't eaten anything for days. In Shangcai, another 600,000 people are surrounded by the flood; 4,000 members of the Liulianyu Brigade of Huabo commune have stripped the trees bare and eaten all the leaves . . . and 300 people in Huangpu commune who had not eaten for six days and seven nights are now consuming dead pigs and other drowned livestock.

August 17: There are still 1.1 million people trapped in the waters, including 500,000 in Shangcai. . . . The disease morbidity rate has soared. According to incomplete statistics, 1.13 million people have contracted illnesses, including 80,000 in Runan and 250,000 in Pingyu; in Wangdui commune alone, 17,000 people out of a total population of 42,000 have fallen ill, and medical staff, despite their best efforts, can only treat 800 cases a day.

August 18: Altogether 880,000 people are surrounded by water in Shangcai and Xincai. Out of 500,000 people in Runan, 320,000 have now been stricken by disease, including 33,000 cases of dysentery, 892 cases of typhoid, 223 of hepatitis, 24,000 of influenza, 3,072 of malaria, 81,000 of enteritis, 18,000 with high fevers, 55,000 with injuries or wounds, 160 poisoned, 75,000 cases of conjunctivitis, and another 27,000 with other illnesses.

August 19: 448,000 people in Zhumadian Prefecture remain in the water.

August 21: A total of 370,000 people are still trapped in the water. . . . Fifty to sixty percent of food supplies parachuted in by air have landed in the water, and 37 members of the Dali Brigade who frantically retrieved and consumed rotten pumpkins from the water have fallen ill with food poisoning.

On August 12, five days after the Banqiao and Shimantan dam collapses, a team of senior officials sent by Beijing and led by Vice Premier Ji Dengkui flew over the devastated area to inspect the damage. Chen Xing, who had slowly worked his way back to prominence after being purged during the Great Leap Forward for predicting precisely the kind of disaster they were now witnessing, accompanied them. In the midst of the massive lake created by the dam failures, the only visible bodies of land were the five county seats which rose like islands out of the lake and were packed with victims huddling together.

The sight of the trapped floodwaters confirmed all of Chen's worst fears. The area, historically a natural flood diversion area, had been subject to such intensive cultivation that eroded soil had filled the rivers and reduced their ability to drain floodwaters. As a result, the floodwaters stagnated above Bantai, and the only way to speed up their runoff was to dynamite the areas blocking water release.

On August 13, Henan Party Secretary Liu Jianxun asked Chen Xing to return to Beijing to report to the State Council on the effects of the disaster. Liu is said to have tearfully requested of Chen: "On my behalf and on behalf of all of Henan, I make just one request—ask them to dynamite the sites blocking the water so that the people of Henan can be rescued."

With the approval of the top leadership, including Vice Premier Li Xiannian and the minister of water resources and electric power, Madame Qian Zhengying, the decision was made to dynamite several of the major surviving dam projects in Henan so that the floodwaters could be released and allowed to drain away. Two days later, under Chen's direction, the offending dams were destroyed, among them the Bantai flood-diversion project whose sluice apertures had earlier been reduced from nine to five in the name of giving "primacy to accumulation." The release of the trapped water created terrific floods down-river on the Huai in Anhui Province.

Fourteen years after the Banqiao and Shimantan disasters, during the debate over the Three Gorges dam, Li Rui cautioned against focusing solely on reservoir storage capacity as a means to achieve flood control. He argued that "in terms of flood protection, dikes are absolutely critical. They are the most effective flood control measures on a river. Their construction and maintenance requires diligent work over the long run, but does not involve grandiose and heroic construction projects. I believe that as long as the earth and rivers exist, dikes will be absolutely necessary."

In November 1975, Madame Qian Zhengying of the Ministry of Water Resources and Electric Power delivered the keynote speech to a national conference on dam and reservoir safety that convened in the city of Zhengzhou, Henan. The following is an excerpt from her speech:

> The "August 1975" storm was a severe test of our work to harness the Huai River and a severe warning about the country's dam projects. We must, therefore, learn from both the negative and positive experiences from this incident in order to improve our overall work proficiency.
>
> Responsibility for the collapse of the Banqiao and Shimantan dams lies with the Ministry of Water Resources and Electric Power, and I personally must shoulder the principal responsibility for what has happened. We did not do a good job: First, and foremost, we have never experienced such large-scale floods or such a catastrophic series of dam collapses. We took for granted that the large dams were safe without researching the issue. We

primarily followed Soviet safety specifications and although we made some design changes, we didn't make fundamental improvements and we didn't draw from the experiences of other countries or from our own experiences. Second, we failed to establish dam specifications in accordance with the unique and special conditions of our own country. Third, we did not do a good job in reservoir management and did not study the problems in any detailed way. There were no clearly designated rules governing reservoir safety, communications, electrical supply sources, emergency procedures, and preparation of necessary materials and equipment. At the most urgent moment after the collapse of the two dams, there was a blackout that caused tremendous chaos. Fourth, there was a failure to establish clear lines of authority and leadership during the crisis. We should be held accountable for that. The two dams played a critical role in electricity generation, but since basic hydrological data was lacking, the design of the flood control system was very flawed and unreliable. After a severe rainstorm in 1972, a reassessment of the flood control system of the Banqiao reservoir was done, but necessary measures were not adopted and the overall safety level of the facility remained relatively low. Owing to the emphasis on water storage and the general ignorance of flood control systems, a 1.9–meter-high earthen ramp was added to the Shimantan reservoir to increase its overall holding capacity. Similarly, before the storm, officials authorized increasing the Banqiao reservoir's capacity by 32 million cubic meters—well in excess of the reservoir's designed safe capacity. Later, in order to protect the lower reaches during a water release, four million cubic meters of water were held back, which hastened the dam collapse. And, as I mentioned above, there was a lack of necessary measures concerning reservoir safety. Electrical power and communication were both cut off during the flood and as a result residents could not be evacuated in time and adequate warnings could not be communicated to residents in the lower reaches.

We must summarize the lessons learned over the past twenty-five years and make adequate and realistic estimates of flood potentials. The 1954 flood* surpassed all our estimates, and in 1974 the floods in Shandong and Anhui provinces were also beyond our expectations. And this year's flood (1975) was no exception. It is said that during the Wanli imperial era during the Ming dynasty (1368–1644), floods of the same magnitude hit the Huai River. We must earnestly summarize the experiences over the past ten years and discern any meteorological or historical patterns.

---

*The 1954 flood devastated the middle and lower reaches of the Yangtze. Thirty-two thousand square kilometers of cultivated land were inundated, 19 million people were displaced, and 30,000 were killed. Previous catastrophic floods had occurred in 1870 and 1949. Further floods and drought in 1956 spurred on supporters of the massive dam- and reservoir-building campaign launched during the Great Leap Forward. Lieberthal and Oksenberg, *Policy Making in China,* p. 272, and Oksenberg, "Policy Formulation."

Qian Zhengying's speech left a deep impression: Dam collapses such as the Banqiao and Shimantan disasters must never recur.

What she failed to say, however, was that [as Chen Xing had pointed out twenty years earlier], the dominant policy of "primacy to accumulation and irrigation" was bound to result in the kinds of disasters that had occurred. She also failed to explain why Chen's ideas were rejected at the time and why he later became the victim of a political purge, only to be brought back after the disasters had struck. On all this, as on the decision-making systems that caused the disasters, she remained silent.

By saying only that she personally shouldered the responsibility, Qian diffused any possible move to pursue more specific responsibility—up to and including criminal legal responsibility—for each and every one of the mistakes that precipitated the disasters. As a result, over the next decade and beyond, the old policy of damming rivers was pursued as blithely as before. And then, in 1993, Lu Youmei [former vice minister of energy and chairman of the Three Gorges Development Corporation] jumped up and proudly claimed that if anything went wrong with the Three Gorges project on the Yangtze, he would be accountable. Lu made the promise while announcing that the date for blocking the Yangtze River [for the Three Gorges dam] would be moved forward a year so as to coincide with the return of Hong Kong in July 1997. This is similar to the situation in 1972 when the completion date of the Gezhouba dam was advanced in order to celebrate the birthday of Chinese Communist Party Chairman Mao Zedong. Such are the effects of personal boasting and bombast.

In ancient times, the harnessing of water by the ancient sage Yu* followed natural laws and respected humankind's desire to survive, raise harvests, and live a good life. He also had great respect for the inherent nature of water. Water and land coexisted peacefully.

The year after the disaster, in early summer 1976, the fertile land where the 85,000 victims were buried produced a bumper crop. Surveying the land carefully, one could see crops everywhere, but what made people's hearts quiver were the small areas where the crops were especially rich and dense.

Looking at the silvery wheat waving in the breeze, one survivor commented: "The wheat is really growing!"

---

*Yu is China's celebrated mythical hero who succeeded the legendary emperors Yao and Shun and who established the tradition of emperors building dams and controlling floods as central to their roles.

Chapter Four

Discussing Population Resettlement with Li Boning

Qi Ren

Author's Note

In early 1992, I was unexpectedly invited to visit the Three Gorges area. Since then I have met with numerous experts and officials familiar with the Three Gorges dam project and have spent a year conducting my own research.

The most troublesome thing about the project is that dam supporters and dam opponents talk past one another, neither side giving serious consideration to the proposals of the other. Worse yet, each side denigrates the views of its opponents, making it difficult for outsiders to gain a good understanding of the project. As a result, problems which might otherwise be solved by a common sense approach have been caught up in the debate and have become increasingly complex. Resettlement is one such issue.

In order to encourage open debate, I have decided to reprint an article by Li Boning in the first part of this chapter. I will follow his remarks with my own.

Li Boning is one of the dam's most vocal proponents. In the mid-1980s, he was the leading candidate for the position of governor of the would-be Three Gorges Province.* He is presently deputy director of the Three Gorges Construction Committee of the State Council, standing deputy director of the Leading Group of the State Council Overseeing Trial Re-

---

*The proposal to create a Three Gorges Province (*Sanxia sheng*) was made in the 1980s but later scuttled. The central government has, however, designated Chongqing as a national municipality which effectively separates the administration of the city from Sichuan Province.

settlement Projects (Guowuyuan sanxia gongcheng yimin shidian gongzuo lingdao xiaozu), *and director of the Economic Development Office of the Three Gorges Area of the State Council.*

Li is generally regarded as the government's most noted expert on resettlement and is the man in charge of the resettlement components of the Three Gorges project. The article that follows systematically lays out Li's views on resettlement, and is taken from his book Developmental Resettlement Is Good.*

Li has reiterated the arguments put forth in this article in speech after speech, and the approach to resettlement he describes has become official government policy for the resettlement of people affected by construction of the Three Gorges dam.

Out of respect for Li Boning I have broken with convention and put his article before mine rather than in an appendix.

General Plan for Population Resettlement

Li Boning, September 1991

The success or failure of resettlement will ultimately determine the success or failure of the Three Gorges project. Both the Central Committee of the Chinese Communist Party and the State Council consider resettlement to be a very important issue and have adopted the policy of "Developmental Resettlement." This new approach to resettlement includes lump-sum reimbursement for lands lost to inundation, rural resettlement, township and factory relocation, personnel training, and other issues pertinent to relocation.

Trial projects have shown that although it is very difficult, resettlement

---

*Li Boning, "Implementation of the New Policy of Developmental Resettlement in the Three Gorges Region," in *Kaifaxing yimin hao* (Developmental Resettlement is Good) (Beijing: Water Conservancy and Electric Power Publishing House, 1991). A portion of this article, including Li's acknowledgments and the dedication of the book is found in Appendix A.

in the Three Gorges area is environmentally feasible. If the Central Committee's resettlement policy is carried out fully and in a timely fashion, it will be possible to move most local populations back from the river and settle them in nearby areas. They need not be moved far away. What follows is an introduction to the issues germane to resettlement and a discussion of trial projects carried out over the past five years.

Problems with Resettlement

When the Three Gorges dam is complete, 19 counties and municipalities will be partially or completely submerged, including two county-level municipalities, 11 county seats, 140 towns, 326 townships, and 1,351 villages. As of 1985, inundation is expected to affect the following:

- Population of areas to be inundated: 725,500
  From townships: 392,900
  From rural areas: 332,600
- Amount of arable land to be inundated: 356,900 mu [23,800 hectares]
  Of rice paddies: 110,700 mu
  Of dry land: 240,200 mu
  Of orange groves: 74,400 mu
- Number of factories to be inundated: 675
  Value of fixed assets of the factories: ¥819 million
- Number of power stations to be inundated: 139
  Total installed capacity: 77,000 KW
- Length of highways and roads to be submerged: 956.1 km
- Length of high-tension wires to be flooded: 1106.4 post [wires] km
- Length of telephone lines lost: 2,729.5 post [wires] km
- Broadcast (nonwireless) communications lines: 5276.2 post [wires] km
- Number of primary cultural antiquity sites flooded: 6

As these figures indicate, the Three Gorges project is unprecedented in Chinese dam-building history. The difficulties faced in construction, especially those related to resettlement, are therefore equally unprecedented.

Population density is very high in the reservoir area, with an average of 1.1 mu of land per capita. There is a lack of irrigation facilities and very little planning for flood prevention. Education levels are quite low. Moreover, during the many decades of uncertainty over the project, there was little economic investment in the region, creating considerable poverty among the locals. Of the 18 million people in the region, more than three million live in poverty, and many counties in the region rely on government subsidies just to get by.

Previous resettlement operations are likely to provide little comfort for potential Three Gorges relocatees. In the past, resettlement was not handled properly; many people were left stranded without employment or adequate shelter and in a state of destitution, creating considerable social and political instability in reservoir areas. These experiences have left people feeling wary of resettlement. Whenever the issue is brought up, people react with fear, and they refuse to discuss why previous efforts failed and how the Three Gorges is different. Once bitten, twice shy.

Certainly, successful resettlement will be difficult and must be taken seriously, but critics' assumptions that resettlement in the Three Gorges area will be a disaster are not credible.

Why the Three Gorges Area Is Well Suited for Resettlement

The Three Gorges area is particularly well suited to accommodate a large resettled population for a number of reasons. The area has abundant natural resources (especially land); most of the resettlers are from townships and are therefore easier to move than are rural people; the project is a long one and will allow sufficient time for proper resettlement; the new Developmental Resettlement policy will ensure that resettlers are well taken care of; and, finally, the central government is firmly committed to the resettlement program and to the project overall.

The Three Gorges area's abundant land resources and burgeoning agriculture, fishing, mining, tourism, and processing and service industries make it particularly well suited to receive settlers. In fact, many of these industries have emerged as a result of the dam's construction. The dam will submerge more than 350,000 mu of land, including 70,000 mu of orange groves, but studies have shown that there is upward of 20 million mu of undeveloped land in the 19 counties and municipalities where people must be resettled, 3.89 million mu of which is found in the 361 townships where resettlement will take place. About 4.2 million mu of the undeveloped land in the area is arable, and 40 percent of it is low-grade sloping land. For each rural relocatee, one-half mu of highly productive land for grain, and one mu of land for orange groves or other cash crops (fruit, tea, mulberry leaves, or herbs) is needed. This means that 500,000 mu of land will be required to accommodate the 300,000 rural resettlers. An additional 30 to 50 percent of this amount must be developed and given to the locals as compensation for lands allocated to resettlers. In total then, 800,000 to one million mu of land will have to be developed to resettle people successfully and assure them adequate grain production, a stable life, and higher incomes than they enjoyed before.

Although it is both possible and feasible to develop the needed land in the 361 townships, if necessary, the areas slated to receive relocatees can be expanded to include other territories within these same counties. It will also be possible to develop some areas that lie below the submersion line. By building dikes to protect arable land in Kaixian, Zigui, Badong, and Wanxian counties, an additional 25,000 mu of farmland can be saved. Moreover, if the dam's normal pool level is reduced to 160 from 175 meters, another 160,000 mu of land can be saved from inundation and used by local farmers to supplement their incomes. Finally, there are also vast grasslands in the Three Gorges area that can be used by relocatees to graze animals and to begin an animal husbandry industry for local markets.

The Three Gorges area also contains abundant stores of natural resources—salt, natural gas, coal, phosphorous, limestone, and marble—the exploitation of which will create thousands of jobs for resettlers. As a start, the government has approved construction of a 60,000–ton-capacity alkaline facility near Wanxian Municipality, which will create 39,000 jobs. Projects like this one not only contribute to job creation and economic growth, they also reduce the cost of resettlement; in effect, killing two birds with one stone. If national and local governments can provide similar opportunities for rural relocatees during the dam's construction, then the cost of resettlement can be reduced substantially. This is in line with the general policy laid out by Comrade Deng Xiaoping, who has said on many occasions that "we should strive to provide more projects in this area," and also with Article 19 of the State Council announcement titled "Land Reimbursement and Resettlement Procedures for Large and Medium-Sized Hydroelectric Projects," which declares that the construction of new production facilities must be combined with resettlement work.

Rural relocatees will be able to continue working in the agricultural sector after they have been resettled. Trial resettlement projects have proven the viability of this approach, and rural relocatees and relevant local governments have guaranteed that the projects will be successful. Finally, where continued work in the agricultural sector is no longer possible, other occupations can be considered, though a large-scale conversion from agricultural to nonagricultural employment will not be necessary.

A second reason why the Three Gorges area is so well suited to receive relocatees is the relative proportion of people being moved from townships as compared with those being moved from rural areas. Just over half of those to be moved (54 percent) are from townships—a much higher percentage than in most resettlement efforts. This is important because resettling people from townships is a simple affair; generally, after moving people work at similar jobs as they did before (and, therefore, do not require retraining), and officials need only expand urban services and functions to the new areas in order to accommodate them. Resettling rural people is more difficult. It often means converting them to an entirely new way of life in fundamentally different occupations.

The rural dwellers who will be moved are scattered throughout 19 counties and municipalities, from Sandouping in Yichang Municipality, to Mudong Township in Ba County. They constitute only 2.6 percent of the total rural population of the Three Gorges area, while the arable land to be flooded constitutes only 2.5 percent of the total. And, significantly, only 2 percent of the rice paddies in the Three Gorges area will be lost to the dam. (Because the region is hilly, much of the arable land is on hillsides above the projected submersion line.)

None of the 326 towns that will be affected by the reservoir will be completely submerged, and only a very small number of villages will be totally inundated. Studies indicate that in 291 of the townships, resettlers will simply have to move to other parts of the town, and in only 35 of them will people have to move to other towns altogether. This is why local conditions are so favorable for carrying out the policy of moving back from the river and settling in nearby areas rather than having to move relocatees over great distances, something which is unique to the Three Gorges project. These studies show that the Three Gorges project will not repeat the problems of previous large-scale hydroelectric projects which required the resettlement of large numbers of people to remote and distant areas.

The fact that the dam will take many years to build is a third factor that will help facilitate successful resettlement. Unlike previous projects in which rivers were diverted quickly, driving people from the area without adequate preparation or planning, the Three Gorges project will give people enough time to adapt to their new environments and employment possibilities.

A fourth factor is the Developmental Resettlement policy, which is designed to provide economic benefits to rural resettlers through the government-financed reclamation of higher-elevation land, the cultivation of cash crops, and the creation of industrial jobs along with lump-sum reimbursements for relocatees' losses. Trial projects over the last five years have proven the success of this approach, and the policy ensures that migrants will be protected by the central government from the start of the project to its finish.

A final factor contributing to the feasibility of resettlement in the Three Gorges area is the central government's firm commitment to the project and resettlement program. Government leaders not only formulated the Developmental Resettlement policy, but also allocated ¥100 million for trial projects over the past five years and formed the Leading Group of the State Council Overseeing Trial Resettlement Projects. None of these steps was taken for previous hydroelectric projects. Moreover, the Economic Development Office of the Three Gorges Project has worked with relevant local governments to coordinate the various aspects of Developmental Resettlement. Together they have made great progress. Over the course of the five-year trial projects, cadres, masses, and migrants have come to appreciate the benefits of the dam and support its speedy construction. Together, the cooperation of the masses, the support of the central government, and the five-year experience with trial projects guarantee that resettlement will be done well.

Still, despite all this evidence that resettlement will be successful, some comrades are concerned that developing more land in the Three Gorges region will contribute to soil erosion and destroy the local environment. This is an issue that deserves attention, but it can be resolved. Strict planning procedures and adherence to stringent quality control standards will be followed in opening lands to grain and citrus fruit tree production. This is not a case of unplanned and haphazard development of barren lands. Indeed, the orange groves which were developed as part of the trial projects required the construction of stone wall terraces [on the mountains] that actually improved soil retention. The same is true where grasses and trees were planted to help in soil conservation.

Trial Projects in Township Resettlement

In planning for new towns over the past five years, we have focused our efforts on building road, water, and electricity projects. These projects, called "three dimensional infrastructure construction projects" (*santong gongcheng*), have created favorable conditions for township resettlement and development.

Through various studies and the experience of the trial projects, we have learned that the long delay in launching the Three Gorges project has had an especially adverse effect on the development of some township economies. Now these towns are so saturated by recent population growth and urban sprawl that there is little room for further development. If the resettlement programs are not implemented soon, given population growth and peoples' desire to escape poverty, the locals may start building projects below the submersion line. In fact, we know that millions of yuan have already been spent on these types of projects in counties all along the river. According to resettlement experts, by 1986, ¥11 billion had been invested in projects below the submersion line. By 1990, the figure was up to ¥18.5 billion and was increasing by an average of ¥1.8 billion per year. If this situation persists, the cost and complexity of successfully resettling people will increase dramatically. Therefore, it is extremely important to carry out the "three dimensional projects" in order to create an environment favorable to investment, to reduce future investments below the submersion line, and to lay a solid foundation for future large-scale resettlement and development.

Over the past few years, we have invested more than ¥9 million in 13 "three dimensional projects," and most are already showing considerable benefits. For instance, Fuling Municipality took advantage of reconstruction work on the Wu River bridge and built a three-level overpass. In three years, 300,000 square meters of new housing and 7.5 kilometers of new roads have been built in the area. Fuling relied primarily on its own resources and funds for the trial projects and received only ¥1.6 million from the government. The projects will greatly reduce the costs incurred by inundation.

Another county received ¥2.1 million for its program to build roads, relocate ten factories, and build an alkaline plant with an annual production capacity of 60,000 tons. Badong County used resettlement funds from the Gezhouba dam and ¥800,000 from the government to help with its project—building 7.7 kilometers of roads and 600 cubic meters of water storage tanks for newly developed areas.

From 1985 to 1990, over ¥62.5 million was invested in "three dimensional infrastructure construction projects" in newly opened areas, as compared to only ¥6.37 million in the older areas of the towns. In other words, over 90 percent of new investment has been devoted to newly developed areas. The investments will reduce the future costs of resettlement and help keep local residents from building below the submersion line.

Factory Relocation Trial Projects

In addition to the "three dimensional projects," we have invested ¥3.8 million of trial project funds to allow factories near or below the submersion line to build new facilities in other locations. For instance, the Badong state-owned cement factory was offered ¥1.3 million (to be repaid after ten years) to build a new factory while the old one was being moved. Because the factory had already been promised ¥2 million in reimbursement funds, this resulted in a ¥700,000 savings. In another case, a county-run nail factory which had originally been promised ¥1.03 million in reimbursement money was instead offered ¥1.5 million in trial project funds to help construct a new facility that is already in production. Finally, a county-run factory received ¥1.5 million in trial project funds to build a new plant. This plan will preempt the necessity of allocating ¥4.5 million to the factory in reimbursement funds.

By relocating factories in this way, we can keep these facilities from expanding below the submersion line, save reimbursement funds, and simplify the relocation process.

Education and Training

Development requires well-trained and well-educated people. Unfortunately, the education level among the people of the Three Gorges region is very low. In the past few years this has become a priority and we have spent ¥1.35 million, primarily for on-the-job training. We have, for instance, offered training for local farmers in the cultivation of oranges; technicians have been trained in various village-level study sessions; and an experimental grove was created where farmers come to learn farming techniques. The farmers who attend these sessions, along with local agronomists are, in turn, passing their knowledge on to other farmers.

The Three Priorities

This chapter has identified Three Priorities which are fundamental for successful resettlement—implementing Developmental Resettlement and opening up new lands to cultivation, building "three dimensional infra-structure construction projects," and educating and training relocatees. To properly implement these priorities, the Three Gorges project must be launched as soon as possible. Otherwise we will repeat previous experiences where migrants were chased from their homes by a reservoir's rising waters. The Three Priorities are reviewed below.

The first priority is in the area of rural resettlement, and specifically land development for orange groves and other cash crops. This is important for the following reasons:

The resettlement of rural people is the most difficult task
facing the project

Our studies show that the relocatees are capable of developing primary agriculture including orange groves and other cash crops as well as animal husbandry, and that adequate space exists for this development. There will be no need to employ migrants in other areas.

To facilitate increased agriculture, land must be developed as quickly as possible. However, the amount of land available is limited, and a variety of interests compete for control over it, including government departments and ministries (forestry, agriculture, water and soil conservation, foreign trade), and orange farmers. The masses have "orange fever," and many of them hope to get rich by growing oranges. But if all of the land in the Three Gorges area is developed by locals looking to cash in on the "fever," nothing will be left for the relocatees and the policy of moving people back from the river and resettling them in nearby areas will not be possible. This would have a devastating impact on the entire Three Gorges project. For these reasons, control over and planning for land use should be put under the authority of the resettlement offices.

The development of between 800,000 and one million mu
of land will provide sufficient land for resettlers and
adequate compensation for original owners

Because orange trees take eight to ten years to mature, work must begin quickly so that resettlers will be able to earn a living once they are moved. If we fail at this task, there will be dire consequences.

Over the past five years, the trial projects have provided valuable experience in primary agriculture, orange groves and other cash crops, introducing a modern style of management, overseeing and controlling income distribution, and carrying out the legal procedures for terminating and

Markers. *(Photo by Richard Hayman)*

and bridges to indicate the height and boundary lines of the areas to be submerged. The markers make it easy for local residents to determine whether they will have to move. But still, experts and officials cannot agree on the number of people.*

According to Li Boning's article, written in September 1991, 725,500 people will have to be resettled to make way for the dam. This was the official figure used later, in April 1992, when the State Council sought approval of the project from the National People's Congress (NPC). However, the 725,500 figure was based on statistics that were compiled in

---

*While the markers do indicate to those living below them that they will have to move, they may also provide for a false sense of security for those living just above them. The resettlement plan includes the resettlement of people living on land below the 177–meter-level, two meters above the normal pool level. It does not include the resettlement of hundreds of thousands of people living between the normal pool level and the crest of the dam (177 to 185 meters). This area is likely to be used as flood control space during serious floods, which are likely to occur about once every 50 years. This fact makes Li Boning's proposal to develop farm land below the submersion line (mentioned earlier in the chapter) even more outlandish. See Philip B. Williams, "Flood Control Analysis," in Margaret Barber and Gráinne Ryder, eds. *Damming the Three Gorges: What Dam-Builders Don't Want You to Know,* 2nd ed. (London and Toronto: Earthscan Publications Ltd., 1993).

1985, seven years earlier. The figure also fails to take into account expected population growth of 1.2 percent per year—something Li did not mention in his article.

In a statement to the New China News Agency three months after his article was written, Li changed his tune and stated that, based on 1990 figures, the dam would require the relocation of 800,000 people. At approximately the same time [late 1991/early 1992], the State Council approved the Assessment Report on Population Resettlement which declared that 1,133,800 people would have to be resettled. [The State Council thereby approved a figure that was substantially higher than the one that it had just submitted to the NPC.] However, even this higher figure was grossly off the mark, and an examination of the assumptions behind it reveals why.

The 1.1 million estimate was based on the assumption that construction of the dam would begin in 1989 and be finished by 2008. However, the vote was taking place in 1992 and the project had yet to be approved, let alone started. Moreover, the delegates were being asked to include the project in the next ten-year plan, which meant that the dam would be launched some time between 1993 and 1999. Given the estimated construction time of twenty years, the dam would not be finished until between 2012 and 2018, and certainly not by 2008. In short, the estimate excluded many years of population growth in the region.

Skeptics of the Three Gorges dam believe that the actual number of people who will be resettled will be much higher than the State Council figure; probably somewhere between 1.3 million and 1.6 million. Zhou Peiyuan, former vice-chairman of the Chinese People's Political Consultative Conference, said the figure would "probably reach 1.6 million," while Li Rui, a long-standing critic of the dam and former vice minister of the Ministry of Water Resources and Electric Power, was more categorical in reaching the same conclusion. "The total *will* reach 1.6 million," he said.

Dam supporters strongly disagree, and Li Boning has sharply criticized these higher estimates. "There are certain people who have not taken part in any of the dam studies who refuse to accept the figures provided by the various resettlement organizations. They claim that the estimate of 1.1 million directly and indirectly affected people is a serious underestimate. Who knows how they came up with their figure of 1.6 million?" Li added that his figures "have been carefully examined and accepted by villages, townships, counties, and districts."

Despite the discrepancy in the figures cited by both sides, I do not believe that Li Boning and the resettlement organizations have ever truly

Table 4.1

Resettlement Estimates

Year plan proposed	Number of relocatees	Time-frame	Source
1985	725,500	Initial calculation	Report submitted for NPC approval
1991	800,000	Recalculation	Official report
1991	1,133,800	Population increase incorporated	Assessment report
1991	1,303,600	Assumes inundation by 2008	Comments by Li Boning at internal meeting
1991	1,314,400	Assumes inundation by 2012	Comments by Li Boning at internal meeting
1992	1,500,000	Assumes inundation by 2000	State Council meeting
1992	1,980,000	Assumes inundation by 2013	Extrapolation from State Council

believed the lower "official" figures despite claiming that they were the result of careful examination. In early 1992, Wang Hanzhang, deputy director of the Hubei Provincial People's Congress and a supporter of the dam, commented to a dozen reporters that "the real number of people to be resettled will reach 1.2 million," and in an informal talk in August 1992, Li Boning also admitted that "the total will surpass 1.2 million."

I was present when Li made this statement. Everyone there was surprised, since it had only been four months since the NPC had been given the 725,500 figure, and less than a year since Li had written the article reprinted above.

I later discovered, however, that even the 1.2 million estimate was not Li Boning's actual working figure. At a restricted meeting on November 6, 1991, Li admitted that the project would require the resettlement of 1,314,400 people. He said: "Statistics from the 19 counties and municipalities [which will be affected by the project] show that population growth between 1985 and 1990 was 2.2 percent. Therefore, by 2008 the number of people to be resettled will not be 1,133,800 but will instead come to 1,303,600. If the project is completed in 2012, the total number of people resettled (according to the Feasibility Report on the Three Gorges Project approved by the Examination Committee of the State Council) will be 1,314,400."* Li's estimates had now surpassed the lower estimates of the dam opponents like Zhou Peiyuan and Li Rui.

So what is the actual figure? I do not have Li's staff or resources, so I cannot provide a definitive answer. But in the summer of 1992, I had the opportunity to participate in a meeting on resettlement convened by the State Council which shed even more light on this difficult question. Also present at the meeting were cadres from the 19 counties and municipalities in Hubei and Sichuan which will be inundated. They had all participated in the dam studies, and I am confident that these were not the people Li had criticized for refusing to accept the official figures. Nevertheless, when they were asked to report on the number of people who would have to be moved from their jurisdictions, their answers were surprising.

The vice governor of Hubei said that by the year 2000, the number of people who will need to be moved from Hubei will have reached 190,000. The vice governor of Sichuan said that by the same year, the total number of people who will need to be moved from his province will be 1.3 million. Together, these two statements suggest that by the year 2000, 1,490,000 people will have been moved. The mayor of Yichang, for his part, commented that by the year 2000, 200,000 people will have to be moved from his jurisdiction. Since Yichang is in Hubei, the vice governor's estimate of 190,000 would appear to be low.** Finally, we must keep in mind that these estimates are for the year 2000, but resettlement is not expected to be finished before 2012. If the estimates are projected to 2012, more than 1.8 million people from Hubei and Sichuan are likely to be moved.

Not only is this estimate more than double Li's first "official" figure, it also surpasses the upper limit of the estimates made by dam opponents. Of course, we should not ignore the tendency in our country of local officials to exaggerate figures in order to receive larger allocations from the central government. But even if the figures are somewhat exaggerated, we can still draw certain conclusions from them.

First, Li Boning and officials in the counties and municipalities in-

---

*According to Li's 1985 proposal, combining the baseline figure for population resettlement and the population growth, the total in 2008 should be somewhere around 1,194,900 and in 2012 it should be 1,303,200.

**Ninety-nine percent of relocatees in Hubei are from the city of Yichang.

volved in resettlement work have never truly "acknowledged" the resettlement figure of 1,133,800 from the 1991 Assessment Report. Second, of the various estimates put forth by dam opponents, the lower figure [1.3 million] is probably closer to the truth. Either way, the number of people who will have to be moved is much larger than the original figure of 725,500 which became the operational figure for the NPC's examination of the Three Gorges issue.

Li Boning and his colleagues have visited the Three Gorges region on numerous occasions, and Li may be more familiar than anyone with the true facts and figures. So, why has he insisted on promoting a lower, untruthful resettlement figure? Zhou Peiyuan believed that this reflected Li's inadequate consideration of various complex factors. At one time I would have agreed, but Li has proven otherwise. After admitting at his informal talk in August 1992 that the total number of people to be moved would be "more than 1.2 million," Li cautioned that "this figure should not be revealed. For now let's just say the total is one million." Perhaps, rather than inadequately considering all the factors involved, as Zhou Peiyuan suggested, Li had instead been very meticulous in examining all of the factors.

In my opinion, Li Boning's reluctance to reveal the real estimates to the public was a tactical decision. Let me cite his book. During the time of his visits to the Three Gorges, Li was the would-be governor of the proposed Three Gorges Province. The tone he used when talking with cadres from the area had a heavy authoritarian flavor. For instance, on December 2, 1986, he commented: "Everyone considers resettlement to be the key issue which will determine whether the Three Gorges project is launched. Therefore, our estimates of the total number of people to be resettled will determine the fate of the project. . . . If, in following the old method, we come up with a large sum for reimbursement, people will be scared away from the project. Therefore, we must be very careful about the figures we cite since they could become a bullet used by opponents."

In the government's view, launching the project was clearly the prime consideration. With that premise in mind, all of the studies and assessments avoided generating resettlement numbers that would undermine this goal. All other issues were made subordinate to launching the project.

Environmental Capacity

The question of whether relocatees can be resettled in nearby areas without destroying the local environment and its capacity to support people is,

I believe, key to the whole issue of resettlement and to the overall success of the Three Gorges project. Optimists, including Li Boning, argue that adequate space exists within the Three Gorges area for resettlement. This was also the final conclusion of the Assessment Report that was approved by the State Council. But pessimists, including Zhou Peiyuan, Li Rui, Fang Zongdai, Wang Shouzhong, Tian Fang, Chen Guojie, and others, all think that the population density of the reservoir area is already too high and that resettling people in nearby areas will lead to overplowing and will destroy the local environment and economy.

Since the establishment of the People's Republic of China over forty years ago, 80,000 reservoirs have been built in China, necessitating the resettlement of over ten million people.*

Most of these people were moved great distances and many were unable to make a decent living after the move. The 1.3 to 1.8 million people who will have to be moved to make way for the Three Gorges dam may suffer the same fate. It is quite possible that there is not enough adequate land for these people in the whole country, let alone in the Three Gorges area. Once we understand this, the so-called "debate" over whether to settle people locally or move them great distances is revealed to be no debate at all. The important issue is how to create an environment that will sustain these people in the reservoir area and not whether this capacity now exists. We know that it does not, but there is no alternative.

The issue is never framed this way. The true purpose of so-called public discussions about resettlement by the scientific community is to take a firm stand that local resettlement is possible. Why else would Li Boning have said that "we must be very careful with the figures we cite since they could become a bullet used by opponents."

In taking this position, however, the optimists have painted themselves into a corner. Of all resources, land is undoubtedly the most important. All industries—farming, forestry, animal husbandry, and fishing—rely on the land. But neither people nor industries can be resettled to castles in the air.

Li Boning claims that infrared aerial photography shows that 3,890,000 mu of barren land is available in the 361 townships that will receive relocatees. He goes on to argue that providing each of the 330,000 rural relocatees with one-half mu of highly productive land for grain production and one mu of orchard land would require a total of 500,000 mu. This,

---

*See, chapters Two, Three, and Eight.

plus the 300,000 to 500,000 mu of new land which will have to be given to local residents as compensation for lost land, will require the development of a total of 800,000 to one million mu of land. This, says Li, is possible and will guarantee both family income and an adequate food supply. When he informed his subordinates of the plan, he spoke casually, indicating that the figure was "only a small part" of the total amount of barren land actually available.

If this were true, then resettlement would be no problem.

However, after a visit to the Three Gorges area, I came away with a very different impression of the land situation from Li's. The area is very mountainous, land is scarce, and population density is high. There is considerable deforestation in the mountains, and most of the arable land is fragmented and spread out over the mountains. While in the area, I examined documents which indicated that mountainous and hilly land makes up about 95.8 percent of the total, and that population density is about 1,000 per square kilometer. There are, in fact, few areas in China with so little available land and such a high population density. For years the locals have been cutting down forests to open new land for settlement, and now 41.5 percent of the land in the area is under cultivation. Vice Premier Zou Jiahua once commented that the struggle between people and land in the Three Gorges area is a tense one. The locals compensate for the lack of adequate land resources by constantly replowing the land in the hopes of eking out two or three crops per year. As a result, most of the relatively barren soil has been overplowed and cannot be further developed. In the past forty years, local forests have been reduced by 50 percent, which in turn caused serious and significant soil erosion. The 40 million tons of rock and earth which flow into the Yangtze annually attest to how quickly the situation is deteriorating.*

Chen Guojie of the Chengdu Geological Research Institute of the Chinese Academy of Sciences believes that even without the dam project and its submersion of vast swaths of arable land the local population already exceeds the environmental capacity of the Three Gorges area by 15 percent. Isn't it strange that Li Boning has apparently never come across these and other figures? Either ignorance is bliss, or he is purposely avoiding the facts.

---

*Other estimates are that as much as 640 million tons of sediment enter the Yangtze annually. See, Vaclav Smil, *The Bad Earth: Environmental Degradation in China* (Armonk, N.Y.: M.E. Sharpe, 1984), p. 87.

Daning River at Wushan—land fragmentation. Water will rise here 60 meters.
(Photo by Richard Hayman)

When I visited the Three Gorges, I asked Yao Binghua, a local resettlement expert, how much land was available for resettlement. Yao is the deputy director of the Resettlement Planning Group of the Three Gorges Project under the State Council and is also a chief engineer and member of Li's group of optimists. Yao has studied the land situation in the Three Gorges area carefully and concluded that there are 3,680,000 mu of undeveloped, barren land in the area. However, he also found that only one-third, or 1.2 million mu of the land, can be developed—a point that was included in the 1988 Assessment Report on Population Resettlement but which Li has never mentioned.

New laws, enacted since 1988, further reduce the amount of land available for development. For instance, the Water and Soil Protection Act restricts development to lands with a gradient slope of less than 25. Before the new law's passage, the limit had been a gradient slope of 30. The law also stipulates that only land with a certain amount of soil can be used for farming. Finally, it bans further development in areas that have already lost more than 30 percent of their vegetative cover. When these factors are taken into account, Yao concludes that only 600,000 to 800,000 mu of the 1.2 million mu can actually be used. These findings are sharply at odds with Li's claim that infrared aerial photography indicates that there are 3,890,000 mu of land available for cultivation. Yao's conclusions were drawn from the same aerial photographs. Clearly, the two men approached the issue from different points of view. Perhaps this example proves that once so-called scientific proof is tainted by the scientist's emotions and subjectivity, it loses all its original scientific value.

But even Yao's significantly lower figures do not account for all of the limits on the development of land in the reservoir area. All of the arable land above the future submersion line is on mountain slopes and is scattered throughout the region. As a result, some of it is unavailable for farming. One experienced local farmer told me that carving 1 mu of arable land out of the steep hills and hanging slopes to create a terraced field actually requires 1.3 to 1.5 mu of barren land. This ratio was later confirmed by the local government. Therefore, the conversion of 800,000 mu of barren land (the upper range of Yao's estimate) would only yield between 530,000 and 610,000 mu of cultivatable land.

Because the majority of the barren land already belongs to the local farmers and not to the government, Li Boning planned on returning a good portion of it to these original owners as compensation for their other losses. Assuming that Li keeps his word, this will further reduce the amount of land available for the relocatees to between 370,000 and 420,000 mu.

In addition to overestimating the amount of land available, Li Boning also underestimates the number of rural resettlers who will require land. According to Li, rural resettlers will make up 45.8 percent of all relocatees, or 332,000 people [of the 725,500 figure]. However, Li has admitted that more than 1.3 million people will have to be moved, 46 percent of whom are rural relocatees, meaning that there will be 602,000 rural relocatees—270,000 more than Li states. Moreover, as we determined above, the actual number of relocatees will be over 1.3 million, and probably between 1.6 million and 1.8 million. Assuming zero population growth over the next twenty years, the number of rural relocatees may actually be between 732,000 and 800,000.

Given the underestimation of the number of relocatees and the overestimation of the amount of land available, Li Boning's guarantee that each relocatee will receive 1 mu of highly productive cash-crop land and one-half mu of highly productive farmland for grain production is an empty promise. Relocatees will, in fact, receive only about 0.6 mu of land per capita, and at the most 1 mu per capita.

One means of dealing with the lack of land would be to convert some rural resettlers to other occupations such as animal husbandry, industry, or commerce. Sources suggest that more than 200,000 people could change occupations; however, the development of industry and commerce in the Three Gorges area is fraught with problems and will not, on its own, solve the main problem of a lack of land.

Another way to increase the amount of land available would be to allow development on land with a gradient slope over 25. Local officials have the power to violate the law and regularly do so in the name of resettlement. When I visited trial resettlement sites with local officials and pointed out to them that many of the sites were built on mountainsides with gradient slopes over 25 (and some over 30), many officials claimed to be ignorant of the new act. Others acknowledged the law, but said that if they obeyed it there would not be enough land for resettlement. Large-scale resettlement operations have not even begun yet. To develop the 800,000 to 1 million mu of land that Li Boning has promised, entire swaths of mountainsides will have to be filled with illegal development, and major conflicts between environment and resettlement officials will inevitably occur.

Even more astounding than Li's claim that sufficient land exists for all resettlers is his belief that the reservoir area can become self-sufficient in grain production, even though it is not currently self-sufficient and the dam has yet to inundate any land. In making this claim, Li makes three questionable assumptions:

First, Li categorically rejects the fact that the reservoir will flood the best local arable land. Instead, he argues that the dam will flood only 2.56 percent of all of the arable land in the 19 counties and municipalities, and only 2 percent of the rice paddies. I have not studied this question thoroughly, but Li's statistics are not borne out by my observations during my visit to the reservoir area. The slope of the land increases as one moves up the mountains, while nearer the river the land is flatter and more amenable to irrigation. Therefore, it is readily apparent that the most highly productive land is closer to the river and below the mountains. Since everything below the 175–meter-line will be submerged, the reservoir will clearly reduce the amount of highly productive land in the area.

Before I had the opportunity to ask anyone for confirmation of my observations, I took a good look at exactly what Li Boning said when promising that the area would become self-sufficient in grain, and everything became crystal clear. He said: "The annual grain output of the 19 counties and municipalities is about 4.8 billion kilograms, or 345 kilograms per mu. The reservoir will submerge 404,000 mu of arable land, which will reduce the annual yield by 161 million kilograms." By simple division, it is clear that the 404,000 mu of land to be submerged are highly productive. According to Li's own figures, the submerged land produces 398.5 kilograms per mu annually—53.5 kilograms per mu more than the average throughout the 19 counties and municipalities. This disproves Li's statement that only a small portion of good land will be submerged.

Li's second assumption is that reforms in hydrological, fertilization, top soil, and plowing methods can increase the productivity of low-yielding sloped land (about 40 percent of land in the area). But Li seems to have forgotten that the majority of the low-yielding land is on gradient slopes over 25. According to the new Water and Soil Protection Act, these lands should be taken out of production.

Li's third assumption is that barren and grassy slopes which cover parts of the mountains can be developed. But even before inundation, the barren and grassy slopes in the area have shown that they are incapable of sustaining cash crops. To convert them to productive land, were it even possible, would be a massive undertaking. A cadre by the name of Lu Chun from the Economic Development Office of the Three Gorges Area under the State Council has concluded that converting each mu of land in the mountains would require 140 explosive charges. Therefore, to convert 800,000 mu of mountainside to cultivatable land would require 1.12 billion such charges, or 15,000 per day for twenty years!

Li has made one final comment which sheds light on his view of the land and resettlement issue in the Three Gorges area. He has said: "We cannot, in the Three Gorges area, adopt the soil and water conservation measures which have been adopted in other areas." From this statement, it would appear that Li is prepared to push aside any and all obstacles to his grandiose plan for the development of barren lands. But the biggest obstacle comes from the law. Will Li require the NPC Standing Committee to grant an easement to the Three Gorges project to violate the Water and Soil Protection Act?

Cost

Both the "optimists" and "pessimists" understand that ¥18.5 billion will not be enough to complete the project. In 1990, officials put the total cost of the project at ¥57 billion. But inflation and other price increases during construction have not been factored into the estimates—what the Chinese call "static" investment. Recent announcements, based on April 1993 costs, put the total cost of the Three Gorges project at ¥95.4 billion, but the resettlement budget has not been adjusted accordingly. Moreover, there have been significant cost increases since April 1993, which have also not been included in the estimates. Assuming that resettlement will be completed in the next twenty years, and assuming that the average inflation rate over the past six years will remain unchanged, the total cost for resettlement will be ¥170 billion.* These facts make me uncomfortable with the term "static" investment.

---

*As of February 1996, only ¥3 billion had been spent on resettlement. See, Xinhua, February 2, 1996.

Chapter Five

The Environmental Impacts of Resettlement in the Three Gorges Project

Chen Guojie

Over the past few years, I have participated in research to assess the impacts of the Three Gorges dam on its local environment. The research has focused on the effects of the project on the Three Gorges reservoir area, the middle reaches of the Yangtze, the mouth of the river, and the sea in and around the estuary of the Yangtze River near Shanghai. The research has not addressed the economic and environmental benefits and costs of the various functions of the dam such as flood control, electricity generation, and navigation. Nor has it examined other potential problems like sediment accumulation and project cost.*

The research shows that although the project will produce some benefits, they will be outweighed by the costs and that although it has certain advantages, it has many more disadvantages. Moreover, the costs and disadvantages of the project will appear almost immediately and persist over the long run, while the project's benefits will not be evident for some time. Perhaps the most important conclusion is that the environment of the Three Gorges area cannot sustain the hundreds of thousands of people who are supposed to be resettled there. If the state does not take this fact into account, the local environment will be destroyed and the future economic development of the Three Gorges area and of the entire Yangtze River Valley will be undermined.

Not everyone agrees that the dam and consequent resettlement will

---

*This article was originally published in *Shuitu baochi tongbao* (Soil Conservation Newsletter), vol 7, no. 5 (October 1987). For a detailed examination of the potential impacts of sedimentation, see Appendix B.

destroy the environment of the Three Gorges region. Some optimistic comrades disagree entirely. Whether these differences stem from different methods of analysis or some other cause, the issue of resettlement requires further study so that colleagues at home and abroad can continue to evaluate and assess the situation.

Respecting Relocatees

Over the past thirty years, the government has shown considerable responsibility in resettling people, but there is much left to learn. In the past, authorities took only a short-term interest in the fate of relocatees. Those who were moved were considered less important than the project itself, and their resettlement secondary to the physical works. The approach was simple: Mobilize the masses with heavy doses of propaganda in support of the project; provide them with reimbursement funds (many of which were embezzled by local government officials before they ever reached the relocatees); and then resettle them. Where differing views or outright resistance to resettlement were found, the government would eliminate all opposition by labeling people "class enemies" and thereby opening them up to possible persecution or exile. As the old Communist Party saying goes: "Once you pay attention to class struggle, everything can be resolved."

The officials generally believed that resettlement was executed efficiently. But it was not. By focusing on the short term, the long-term social and economic development of the resettlement areas was ignored. Moreover, the relocatees had very little say about the resettlement process, such as conditions in the resettlement areas, how their basic needs should be met, or how resettlement should be financed and managed. As a result, many state-sponsored projects have left a plethora of unresolved problems in their wake.*

If we rely on the old approach to resettlement for moving people out of the Three Gorges area—issuing decrees to move people, reimbursing them for lost land and housing stock at far below market value, and concentrating solely on the project without helping to develop the resettle-

---

*Examples include the Sanmenxia dam on the Yellow River, the Danjiangkou project, the Wujiangdu River reservoir, and the five large-scale reservoirs built in western Anhui Province which drove 100,000 residents out of the Dabie mountain region and left them homeless. See Chapter Eight for a discussion of resettlement in the Xin'an River Power Station Project.

ment area—the same problems will occur. It will not be easy to fully respect the rights of the relocatees and show a true sense of responsibility toward them. In fact, it will complicate and draw out the project considerably. When the organizational departments in charge of resettlement recognize this fact and feel that their task is complex and fraught with difficulties, I will be at ease. If they treat resettlement lightly and feel their task is a simple one, I will be extremely concerned for the fate of the relocatees.

Resettlement and the Environment

Resettlement is an environmental issue. The success or failure of the resettlement program will be determined not only by whether relocatees have food, clothes, shelter, and employment, but also by whether the environment can sustain the incoming population, and whether there are adequate resources available for economic development. If the relocatee population strains or surpasses the region's environmental capacity, the local environment will deteriorate, natural resources will dwindle, and the standard of living of relocatees will fall. The fact is that the environmental capacity of the Three Gorges area is already strained and the resettlement of relocatees in the region will only make the situation worse. Some serious concerns include population pressures and consequent overplowing, deforestation and soil erosion, and the effects of the project on the rural labor situation.

Population Pressure and Overplowing

To describe the Three Gorges region as "underdeveloped" is misleading. It is, instead, maldeveloped. To meet the needs of the present population, local forests have been destroyed and a great deal of steep land has been converted to terraced fields for cultivation. Most of the land in the area (78 percent) is mountainous, and about 40 percent of it is under cultivation. A third of this land is on mountainsides with gradient slopes of 25 or greater and therefore should not be farmed (according to the Water and Soil Protection Act). Because of the region's growing population, the average resident has only 0.07 hectares of land and produces only 340.5 jin* of

---

*One jin equals one-half kilogram.

grain—far below the national average and the averages for Sichuan and Jiangxi provinces.

Deforestation and Soil Erosion

There has been a dramatic loss of forest cover in the Three Gorges area. In the 1950s, 20 percent of the area was forested, but today there is only 10 percent tree cover. In Fuling, Fengdu, Kaixian, Fengjie, and Zigui counties 95 percent of the remaining trees are immature. Horsetail pine makes up more than 70 percent of the trees, and there are only about 4.35 square meters of trees per hectare. These areas are regressing from forest to bushes to grassy slopes and finally to exposed rock. Over 80 percent of the land is experiencing soil erosion, and more than half of it is serious or extremely serious. Soil erosion causes about 40 million tons of sediment to flow into the Yangtze each year.

The deforestation and erosion also cause landslides. There are 172 gullies, 210 landslips of more than 600,000 cubic meters, and 36 large-scale landslips of more than 10 million cubic meters. The sediment problem is so serious that in one county 60 percent of the storage capacity of its 2,200 reservoirs has been lost to accumulated sediment.

Rural Labor Surplus

For the sake of the population and the development of the reservoir, one must discuss the issue of exporting labor out of the region. Taking into consideration the fact that much good quality land in the reservoir area will be submerged and that China's Forestry Protection Act (*Senlinfa*) requires vegetation on 40 percent of land before it can be cultivated, forest land in the area [with forest coverage of less than 40 percent] should be protected from cultivation, all of which will compound the problem of surplus population. Thus, it is impossible to resettle large numbers of relocatees in this area.

Ideal Versus Reality

Some residents of the reservoir area are hoping that resettlement funds will radically change their lives for the better—they dream of new towns replacing old ones, of shabby houses being torn down and replaced with

Recent landslides in Xiling Gorge. *(Photo by Richard Hayman)*

modern ones, and of lucrative industry jobs replacing agriculture. But this is highly unlikely. The fact of the matter is that the state does not have enough money to complete the project, let alone the resettlement aspects of it, and will have to rely heavily on borrowed money.

The Three Gorges project's original design was for a smaller dam, with a 150–meter normal pool level, 25 meters lower than the 175–meter normal pool level design that was eventually agreed on. The lower dam would have caused only ten towns to be submerged and therefore would have incurred much lower resettlement costs. Based on this lower design, and using the lowest available estimates of per capita income at the end of the century as a guide, the Urban Planning Research Institute estimated that resettlement would cost ¥500 million. However, even this conservative estimate is substantially higher than the ¥110 million the Yangtze Valley Planning Office plans to provide for resettlement for the now larger dam. It is clear, therefore, that there is unlikely to be enough money to meet the basic needs of the relocatees, let alone make them rich.

Whatever benefits the project brings will not be realized for at least twenty years. But the costs will be felt much sooner. Soon after the project is launched it will wreak havoc on the systems of production. With the old system destroyed and the new one not yet functioning, chaos will ensue. The losses will be immense, and whatever resettlement funds are available will do little to remedy the situation. Experience has shown that even after acquiring electricity, farmers who have lost their land to a reservoir remain poor. In some past projects, enraged farmers have demanded that reservoirs be destroyed and the land reclaimed. We must never forget these lessons.

Past resettlement experience also shows how idealistic plans and promises for compensation and future development often sound plausible, but turn out to be impractical or even harmful when implemented. This same idealism is evident in resettlement planning for the Three Gorges area. For instance, there are those who feel that salt reserves in the Three Gorges area are so vast that once developed they will provide jobs for many relocatees. The reality, however, is not so simple. The salt is buried deep in the ground and would not be easily extracted. Salt mines can be highly polluting, and the Three Gorges area is steep and mountainous with a poor transportation system. All of these factors would make it very difficult for mines in the Three Gorges area to compete with those in Zigong and Lushan in Sichuan. Finally, the Three Gorges region is a renown tourist destination, and industrial development [and consequent pollution] could adversely affect the tourist trade.

Successful resettlement in the Three Gorges area requires a thorough study of every aspect of the dam. Plans for local development and for resettlement could then be tailored to fit the specific needs of the project; to build on its strengths and address its weaknesses. Planning for resettlement in any other way will, ultimately, be unsuccessful. A key element in this planning, and one that can truly increase the wealth of local people, is education. Experience the world over demonstrates that education is the best way to increase economic prosperity. We must also, however, focus on construction, environmental cleanups, protecting forests and increasing forest cover, reducing the rate of population growth and managing population density, adopting long-range economic development plans, improving labor skills, developing primary agriculture and specialized products, and integrating the local economy with that of the region.

Examining the twists and turns in national policy over the last few decades in our country [from the Great Leap Forward to the Cultural Revolution], we see that the Chinese people have made great strides but have suffered as well. We are still poor and we have many lessons to learn. We have made mistakes, including unbridled population growth in the 1950s, poor management of natural resources, and the fostering of environmentally destructive enterprises. We have also made mistakes in macro-level decision making where large-scale construction projects were concerned. The Three Gorges project could be an error of great magnitude because it touches on all of these past errors. Hopefully, the final assessment of the project will be based on scientific fact and we will not repeat past mistakes. We should not be blamed for having such hopes.

Chapter Six

What Are the Three Gorges Resettlers Thinking?

Ding Qigang

As construction of the Three Gorges dam project begins, the primary problem confronting dam builders is population resettlement. At Sandouping [Sandou Township], where roughly 100 families have to be relocated, local officials in charge of resettlement thought that as long as the new site in the city of Yichang was better than people's present homes they would gladly move. But, to the officials' surprise, resettlers from Sandouping have refused to move.

The officials were dumbfounded. "Had the government promised more than it could deliver?" "Were the resettlers trying to bargain for a better deal?" The questions arose out of a belief that successful resettlement is simply a matter of economics and government planning. If relocatees are financially compensated for submerged lands and provided with the means to make a living in their new location, then, the officials believed, resettlement should not pose any serious problems.

But resettlement is not simply a function of compensation and planning. It is a matter of adaptation by and rehabilitation of relocatees. The people from Sandouping refused to move to Yichang because officials would do nothing to help them adapt to their new situation. Of course, economic considerations and government planning can help people adapt, but the characteristics of the resettlement sites—the type of land and the social make-up of the area, for instance—loom larger than economic considerations and coercive government action in determining the success of resettlement. Because the people to be resettled for the Three Gorges dam project are being moved against their will, the success of the relocation program will hinge on how well people settle in to their new surroundings and adapt to new economic, social, and natural realities.

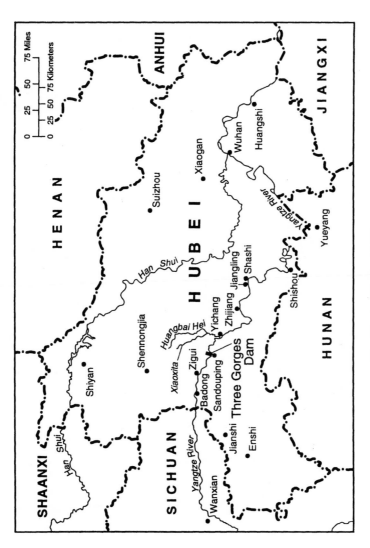

The Three Gorges region.

With this in mind, in 1992 my research team visited the Three Gorges area and conducted fieldwork in three villages: Dongrangkou, located on the northern bank of the Yangtze across from the city of Badong in Hubei Province;* Yangjiapeng, located on the southern bank of the Yangtze about 47 li** east of Badong; and Luoping, situated on the upper reaches of the Longchuan River, a tributary off the northern bank of the Yangtze. Residents from each of these villages will be resettled, though the nature of the move will differ for each community. For example, it is impossible for Dongrangkou to be moved back from the river and resettled in a nearby area like most of the villages in the valley. Yangjiapeng Village will have part of its population moved back from the river and settled into nearby areas, but half will be moved to other villages. Finally, villagers from Luoping will be moved back from the river and resettled in nearby mountain areas, but there is not enough land for about 100 of Luoping's households.

We systematically surveyed 100 of the 570 households in Luoping and Yangjiapeng villages and gathered additional data from Dongrangkou. Information was gathered by questionnaires and supplemented by in-depth follow-up interviews.

Awareness of the Proposed Functions of the Dam and Potential Problems

Generally, people's awareness of the project is related to their understanding of the difficulties involved in its construction. This level of understanding is, in turn, related to people's willingness to cooperate with the government's plans and to the sacrifices they are willing to make in the name of the project.

Our research found that 72 percent of the villagers sampled were aware of the dam and the proposed construction site at Sandouping, while the other 28 percent had heard of the dam but were unaware of its planned location. More than half of the respondents (52 percent) learned about the

---

*Badong has been a center of munitions production since the 1960s when China moved its defense industries into the interior in anticipation of foreign invasion. Recently, it received substantial investments from the central government to support resettlement and reconstruction of the town above the proposed water line of the reservoir.
**One li equals one-half kilometer.

Badong City. *(Photo by Richard Hayman)*

dam from television and other broadcast media, while 32 percent heard about it "through the grapevine," and 16 percent read about it in official documents and announcements. When asked "what is the purpose of the Three Gorges dam?" 70 percent of respondents showed at least some understanding of its function: Ninety-eight percent of these identified electricity generation as the project's primary function and were aware of the country's electricity needs; 21 percent answered that the dam would contribute to flood control; 17 percent said that the dam would facilitate river navigation; and 7 percent believed it would strengthen national defense.

When asked whether the central government faces any difficulties in the construction of the dam, 55 resettlers answered in the affirmative, 23 did not know, and 7 said that there were no difficulties. When asked about the specific difficulties the project is likely to encounter, 80 percent (of the 55) identified the cost of the project as a potential problem, while 20 percent felt that resettlement could be a major difficulty, and 15 percent identified other technical difficulties.

Follow-up interviews indicated that the mass media played an important role in helping relocatees understand the various problems involved in the dam's construction. Respondents felt that newspaper and television reports along with trial resettlement projects have informed public opinion and created an ideal image of the project, in addition to helping people prepare for resettlement. "Even those unwilling to move now believe that resettlement is inevitable," they asserted.

People's attitudes toward the project indicate that the media and the government are having an impact in shaping popular opinions about the project. The most important focus of the propaganda campaign is on electricity generation. The Three Gorges area is one of mountains, reservoirs, and dams, but its economy is underdeveloped, with an extremely weak industrial base and inadequate supply of electricity.* Electricity generation is, therefore, attractive to relocatees who feel that the new power will help develop industry in the area.

When interviewed, nearly half of the relocatees said that they felt the central government would have no difficulty building the dam. "The central government would never have planned the dam if there were many difficulties," was a typical comment from respondents. Nor would the government "start a project it could not complete." But people may not be as optimistic about the project as they first appear. The average annual income in Badong County is less than ¥400, but many people have claimed that their income is as high as ¥2,400 or ¥3,000. The hope, of course, is that by claiming higher incomes they will receive more compensation under the resettlement plan. In other words, relocatees may minimize the difficulties involved in the dam project, while also overstating the value of their assets in the hope of profiting personally from the dam's construction.

Attitudes About Possible Impacts of the Dam

For people to make informed decisions about whether they will move, they must have a good understanding of how the dam will affect their lives and livelihoods.

When asked how the dam will affect their living standards, 22 percent answered that one benefit from the dam would be improved transportation along the river.** The majority of those responding in this way came from families who earned at least part of their incomes from navigating the

---

*The shortage of electricity in China, especially in rural areas, is readily apparent because of periodic stoppages (*tingdian*) that afflict industry and agriculture alike.

**Dam supporters argue that transportation costs will drop by an estimated 35 percent following completion of the project.

Table 6.1

Expected Benefits of the Three Gorges Dam

Benefits expected	Percent
Improved river navigation	22
Electrical generation	10
Regional economic development	8
Greater opportunities for relocatees	2
Unclear	33
See no positive benefits	22

N = 100 Households

river. Other benefits identified by the respondents included electricity generation for the villages, economic development of mountainous areas, and new opportunities for relocatees. Fifty-five percent were unsure of prospective benefits, or felt there would be none.

In the interviews, relocatees emphasized that before the dam had finally been approved by the central government in 1992 their villages were starved for investment. Only after the project began did county authorities get involved in local economic development. Up to that point, farmers had been left to develop uncultivated land on their own. But with the project under way, land development is now the government's responsibility. For centuries, villagers had to fetch water by carrying it up from the river themselves. Now the government has started to fund projects to improve access to water supplies. Without the Three Gorges project, many villagers felt, there would be no investment in their villages.

Among those surveyed, there was considerable agreement about the possible negative impacts of the project. Fifty-two percent of respondents said that the submersion of homes and orange groves was the most serious of the negative impacts, while significant numbers argued that the difficulty of finding suitable agricultural land in the mountains was also a serious concern.

In the interviews, many of the resettlers expressed concern about the inundation. "The losses stemming from inundation will be tremendous," many respondents commented. "Most of our land and houses—built over the past two-hundred years by a dozen generations—will be submerged. We

Table 6.2

Potential Negative Impacts of the Three Gorges Dam

Negative impacts	Percent
Submersion of housing and/or fruit trees	52
Inability to move back from the river and resettle in nearby areas due to overpopulation	14
Problems of daily living	18
Inability to collect sand	2
Unclear	12
No negative effects	2

N = 100 Households

will also lose the orange groves we have been cultivating since the 1970s—our sole source of cash income." Many interviewees also felt that all of the mountainous land suitable for agriculture had already been developed and that population pressures were already placing undue stress on the land. The ¥550 per mu in compensation for lost land was not enough, they believed, to convert steep mountain land into cultivatable terraces.

Attitudes About Resettlement

Are villagers willing to resettle? Sixty-seven percent of those we interviewed claimed that they would be "willing to move in the national interest," while 33 percent were completely unwilling to resettle.

Follow-up interviews indicated that villagers' willingness to resettle is dependent on the perceived impacts of relocation on their families and standards of living. All indications suggest that after inundation most rural resettlers will be left to cultivate inferior land and that their standards of living will indeed fall.

Yangjiapeng Village

Some of Yangjiapeng's residents will be moved back from the river and into nearby areas, while others will be moved to villages far away. Yangjiapeng is a hilly village whose best agricultural land is near the

Village site similar to Yangjiapeng. *(Photo by Richard Hayman)*

river. This good land will be completely submerged by the dam's reservoir, and less than one-third of the moderately sloped dry land now under cultivation will be retained. Of the new land being developed for resettlers, some is 600 meters above sea level with only about 20 to 30 centimeters of topsoil. The land is steep and covered with crushed stone. In addition, land with a gradient slope of more than 40 cannot be converted into terraces. Even if it could be, there are no rocks large enough to build the terrace walls. And, finally, terraces built on such steep slopes are routinely washed away in severe storms. To quote the locals: "This type of land is prone to both drought and erosion. Three moons without rain will dry up the land, and the first rains will wash it all away." The situation is compounded by the lack of irrigation facilities in the mountains. The available spring water is sufficient to meet human needs, but is insufficient for large-scale agriculture. Time and again, local villagers told us that 4 or 5 mu of the proposed newly developed land cannot compare to 1 mu of river valley land.

Luoping Village

The residents of Luoping Village face a similar situation. Surrounded by mountains, the village sits on the banks of the Longchuan River, a tributary of the Yangtze River. The river is the lifeblood of the village. Residents wash, bathe, fish, and irrigate their crops with water from the Longchuan. Yearly floods deposit sediment on the banks of the river, providing much needed nutrients for the soil and sustaining village agriculture. Once the dam is built, however, all of the land below the designated reservoir line will be submerged and local residents will be resettled up into the mountains. There they will struggle to raise crops on steep mountain land.

Over the past eight years, the government has invested about ¥500,000 to develop more than 1,000 mu of land in the mountain areas near Luoping Village. But it is not enough. There are still 100 households that will have no place to go when the waters rise behind the dam. According to the locals we interviewed: "Not only can the new land not support many people, it will require a great deal more work to manage." To complicate matters, many villagers will have to take a ferry across the two–kilometer–wide lake that the dam will create just to reach their agricultural plots.

Our investigations revealed that the percentages of villagers willing and unwilling to resettle are more or less the same in all three villages. Those farmers who will not lose their lands to the reservoir have no reason to move. Currently, their lands are nearer the mountains and of lower quality than river valley land, and consequently, their incomes are lower than those of people now living on the river. Once the dam is built, however, the waters will rise to their doorsteps, offering a host of new economic opportunities. The poorest land today will become the most valuable land tomorrow. By contrast, those villagers whose land and houses will be submerged will witness a dramatic drop in living standards because they will have to move from their highly prized land holdings near the river to undeveloped lands in the mountains. Given the prevailing view that these new lands "cannot support people," the farmers are generally unwilling to move.

Concerns About the Policy of Moving Back from the River and Resettling in Nearby Areas

Resettlement is difficult and uncertain regardless of whether it entails moving back from the river and settling in nearby areas, or moving much further away to entirely new locations. When asked about the official policy of moving people back from the river and settling them in nearby areas, 64 percent thought that it would be both difficult to find suitable sites and prohibitively expensive to build new homes. Twenty-seven percent felt that there was not enough good quality land in the new area for everyone. "Hilly, steep land will not grow grain," is a common refrain. Another 9 percent were concerned that they did not have the manpower to make a successful move and build a new house.

The survey responses indicate that the villagers' primary concern is housing, with the availability of new land an important second. The villagers' houses are of different styles, sizes, and ages, and so their values (for compensation) should also be different. But the government's compensation policy is the same for everyone—¥55 per square meter. In our interviews, the villagers made it clear that they would not accept such a low price. As one farmer put it: "When I first built my house, I took out a loan that I have yet to repay. I don't have the money to build a new house." Another commented: "Before, building a new house was cheap because relatives and neighbors would help with the construction. Now it's different. Each laborer must be paid ¥5. I simply can't afford it." Yet another said: "We'll have to build a new house prior to dismantling the old, and so we can't reuse the materials from the old house. The ¥55 per square meter is not enough to buy materials for a new house."

Generally speaking, villagers think that the government should reimburse them for any and all losses of housing stock, although a few are more concerned that the government provide them with new land since their current land holdings will be inundated. If the government is unable to provide new land, these people argue, it should make sure that they secure nonagricultural employment. In any event, villagers are generally less anxious about the amount of land they are to receive than about reimbursement for their houses. They generally believe that the government will provide some kind of gainful employment and will not allow them to go hungry, but there are no such guarantees for adequate housing.* Village cadres agree that unless the housing problem is solved, it will be very difficult to properly implement the resettlement policy.

The Importance of Community and the Quality of Housing

According to our survey, 77 percent of villagers would prefer that their entire village be moved as a unit. Nineteen percent are willing to accept

---

*Indeed, "reservoir relocatees" from previous projects have been left to languish in temporary housing for years. See Chapter Seven.

whatever form of resettlement the government chooses, and only 4 percent want to be integrated into a new village.

When asked, in follow-up interviews, why they favor collective resettlement, people emphasized the importance of community. In Chinese culture, familiarity and community are of the utmost importance. A popular saying among villagers is that "one friend is better than three strangers." If villagers do not know someone well, communication is difficult, trust is not forthcoming, favors will not be returned, conflicts are likely, and acceptance into a new community is all but impossible. It is because of this difficulty in integrating into unfamiliar social surroundings that most villagers would like to be moved as a group.

If villagers cannot be moved as a group, most want to move to as familiar a setting as possible. Forty-one percent said they would prefer to live with someone from their village, and 37 percent would like to live with their current next-door neighbor. As the Chinese saying goes: "Nearby neighbors are better than distant relatives." Thirteen percent said they would like to live with relatives, and only 9 percent said they had never considered the issue.

Clearly, villagers consider their fellow villagers, old neighbors, relatives, and friends as familiar. Social as well as cosanguineous and marital ties constitute the major social contacts in the daily lives of the villagers and reinforce the distinctions between "we" and "they."

Those villagers who are willing to move have standards regarding their new homes that they are loathe to compromise on. Most importantly, they insist that the new home be at least as good as their original one. Relocatees are also concerned about a whole host of other factors: Eighty-two percent of the people are concerned about whether they will be able to sustain their present standard of living; 73 percent insist on moving to a place that is near the river and near roads; 69 percent want to live in an area with a geography and climate similar to that of their present home; 39 percent want the government to handle everything for them, including resettlement and securing new jobs; 38 percent insist on moving to a place closer to a town; and 20 percent insist on moving to a place close to where they can intermingle with members of their clan.

Opinions About Agriculture

Farmers in the Three Gorges area have developed dry-land agricultural skills suitable to the local climate; skills which are based on thousands of years of experience. Large-scale agriculture is not practiced in the valley. (In fact, tractors are used mostly for transportation and not for plowing fields.)

The most common agricultural implements are hoes and scythes, which are used to cut wheat, weed, and dig sweet potatoes. Major dry-land crops include sweet potatoes, millet, and wheat, while the more humid climate of the valley produces semitropical crops including bananas, palm trees, and sisal hemp. Because local transportation is underdeveloped, most of the produce is consumed locally and there are few cash crops, except for oranges.

Since the early 1970s, orange groves have proliferated in the region. They are a viable cash crop because oranges can be stored for long periods of time and can withstand long-distance shipment. Virtually every household in our survey owns an orange grove, and 37 percent of the farmers said they had been trained to grow oranges by the county or township government. And the vast majority of those we asked (80 percent) said that they would like to continue to grow oranges after they are moved to make way for the dam.

Nonagricultural Employment

Some villagers are willing to convert to nonagricultural employment after being resettled, but a number of factors affect this choice, including age, gender, level of education, and traditional ways of making a living. Age and education also affect employers' decisions when hiring help. Those over the age of forty and with little education will have a hard time finding jobs outside of agriculture. Women, for their part, will also find it difficult to get jobs in industry, transportation, and mining and are likely to be limited to commerce, service industries, and food services.

Our survey indicates that 87 percent of the villagers are willing to convert to nonagricultural employment. Young people are most willing to make the conversion, those over forty least so.

Among those willing to convert to nonagricultural employment, 47 percent said they would like to work in the transportation field because they have a family history of piloting ships along the river or of driving vehicles. Forty-four percent said they were willing to work in commerce or the service sector and most of these already have experience in these fields. Finally, 45 percent were interested in factory work, where they would learn new skills.

Table 6.3

Expectations for Family Income Following Resettlement Among Residents of Yangjiapeng Village

Expectation	Percent
Drastic decline (by 50 percent)	42
Depends on government policy	26
Slight increase	10
No opinion	22

  N = 50 households

Expectations for Family Income Following Resettlement

Yangjiapeng Village

Inundation will force all of Yangjiapeng's residents to be resettled. Two-hundred mu of new land has been developed near the village, but there is not likely to be any further land development. To date, no one knows who will be forced to move afar and who will be able to remain nearby.

If people are moved back from the river and up into the mountains to make their livings in agriculture, what are the anticipated effects on family income? Forty-two percent of the 50 households we surveyed expect their incomes to drop drastically, perhaps by as much as 50 percent. This expectation results from the fact that many farmers actually own much more land than is registered with the village. When the Agricultural Responsibility System was adopted in 1978, the farmers converted as much nearby barren land to farming as was possible.* The converted land does not need to be registered with the village; nor is it taxed by the government. In effect, it is a source of invisible income for the farmers. However, the government will only compensate farmers for registered land lost to the reservoir and, as a result, most farmers will see a drastic drop in both their land allotment and their incomes.

---

*The Agricultural Responsibility System dismantled the collective agricultural system of People's Communes and granted farmers the right to lease land for fifteen years or longer.

Aside from having smaller land holdings after inundation than before, many farmers feel that the poor quality of the new land is the most serious threat to their income earning potential. "Five mu of new land in the mountains is not worth one mu of land near the river," say the farmers. The new land is too steep, its topsoil is too shallow, and irrigation facilities are scarce, making it difficult to grow grain and commercial agricultural products. Moreover, the land is 600 meters above sea level, and oranges grown at this altitude will be inferior to those grown near the river. No matter how much is invested in the land, the farmers say, it cannot be made as highly productive as their current holdings. Twenty-two percent of the farmers find it difficult to anticipate the effects of resettlement on family income. They feel that the surest way to avoid a life of hardship after resettlement is to have household incomes drawn from both agricultural and industrial jobs.

Luoping Village

For the residents of Luoping Village who will have to be resettled, new homes built on steep mountainsides will have none of the advantages of the low-land agriculture to which they are accustomed. Moreover, 400 people or about 100 households will be left with no land at all.

Thirty-eight percent of the 50 households we surveyed in Luoping Village feel that their incomes will drop drastically following resettlement, mostly because of the inferiority of mountain land. As with Yangjiapeng villagers, 24 percent of those from Luoping feel that success or failure rests on the shoulders of the government. They think that incomes will increase if nonagricultural opportunities are offered, and decline if such opportunities are not available. Twenty-two percent feel that their incomes will improve slightly because they will be able to start businesses in the new town or take advantage of the new reservoir to engage in water transportation. Ten percent think that their incomes will improve dramatically because a socialist country must take responsibility for people slated for resettlement and because the central government has promised the local county that arrangements will be well planned.

Overall, Luoping residents are slightly more optimistic about future incomes than their counterparts in Yangjiapeng. While the natural conditions in both villages will decline after inundation, Luoping will be left with land similar in quality to that presently enjoyed by Yangjiapeng

Table 6.4

Expectation of Family Income Following Resettlement Among Residents of Luoping Village

Expectation	Percent
Drastic decline (by 50 percent)	38
Depends on government policy	24
Slight increase	22
Dramatic increase	10
No opinion	6

N = 50 households

residents, while Yangjiapeng's new land will be substantially worse. The decision to resettle Luoping as one group to a single area has boosted the confidence of the villagers; however, those who anticipate drastic improvements in income should beware.

The government's propaganda campaign has fueled people's desire for compensation. A variety of slogans have misled some villagers into assuming that resettlement will make them rich. A popular local saying is indicative: "Thinking about the Three Gorges we look forward to earning big bucks." It should be pointed out as an aside that similar tendencies have been evident among China's transient workers and floating population.* Well-connected people in the Three Gorges area have tried everything they can to move their relatives' registry (*hukou*) from noninundated areas to inundated areas and to places from which resettlement is necessary, all in the hopes of receiving generous government compensation.

The propaganda gives villagers the impression that they can make a fortune from resettlement and that drastic increases in income will inevitably follow. Local county governments are presently trying to counter the propaganda and inform people that the project will not be a cash-cow, but villagers are less interested in this new message than in the old promises of wealth. While unrealistic promises might make it easier to move people

---

*This group is estimated to number as many as 100 million people who move from rural to urban areas and from city to city in search of employment and quick money.

initially, broken promises will make the process much more difficult in the long run.

Expectations of Villagers Willing to Be Moved Great Distances

The entire villages of Yangjiapeng and Luoping must be moved. Owing to land scarcity, many resettlers will have to be moved to different villages, townships, and even counties. Of the 77 relocatees we interviewed who want to be resettled collectively, 38 will agree only to being resettled nearby, while 22 will accept having to move further away. Either way, villagers have similar expectations about resettlement. These can be broken down into four basic expectations or demands.

"Refuse to Move Up into the Mountains"

People living near the river get visibly upset whenever moving up into the mountains is mentioned. Some are so categorically opposed to resettlement that they have vowed never to move, even if it means "doing without land or housing and living by theft and robbery." Such an adamant attitude is surprising, but people understand that life in the mountains is hard and they are simply not interested. One village head said: "Life in the mountains is not half as good as life on the river. Transportation and trade are impossible, and it's futile to try to raise oranges there." Others said: "Life in the mountains is difficult. We are accustomed to living by the river, and we're comfortable with the climate here. It's too cold in the mountains." Near Yangjiapeng there are two villages named Yazi and Xiongjiachong which are 800 meters higher in elevation. People from those two villages belong to the same Tian family clan as people living in Yangjiapeng and readily admit that life on the river is far superior to life in the mountains.

Previous attempts to resettle people from Yangjiapeng have ended in failure. In the early 1950s, the local government moved 60 families to the high mountains of Nanhangling. New houses were built, and the villagers were expected to settle down happily. But before long the families moved back down to the river. A similar situation occurred in the 1960s, when 30 families from Dongrangkou Village were moved to a high mountain plateau 1,800 meters above sea level. Within ten years, they had all returned to their previous homes and occupations. In the early 1960s, the Badong County government established a state farm and moved some people to

Dahuaping, which was also located deep in the mountains, but this too was a complete failure. One after the other, farm workers returned to their homes leaving only the farm director at the site. The farm was literally destroyed by the people.

"Never Leave the River"

Villagers on the river have made a living there for generations and have developed a way of life intimately related to water. Their lifestyle is completely different from that of mountain dwellers. Although the Gezhouba dam has blocked fish from traveling from the lower reaches to the upper reaches of the river to spawn, and despite the effects of overfishing, most people can earn at least ¥8 to ¥10 a day by fishing. Whenever there is no work in the fields, villagers go to the river to fish. They choose a place along the river bank and scoop up their nets every ten seconds or so until they catch something. Those who have money buy fishing poles and line them up along the banks of the small streams that empty into the Yangtze. Despite limited fish resources, fishing provides a steady source of income for the villagers throughout the year. If they catch a small fish, they'll take it home to cook. If they catch a larger one, they take it to a market to sell. And if they catch a really large fish weighing more than a jin, they'll send it directly to the county guest house where it will fetch ¥30 per jin.

Life along the river has also meant easy access to convenient water transportation. The rapid currents of the Yangtze are filled with twists and turns which can capsize boats easily. This difficult environment encouraged villagers to develop their piloting skills. Today, piloting boats is a long-held tradition for many families. For instance, the Tian family has piloted boats for over two-hundred years, and, at one time, owned over a dozen craft. Villagers who do not own their own boats work on others. The residents of Yangjiapeng Village own 11 boats, the largest of which is 40 tons. For many, shipping is the primary source of income.

"Remain Near Towns with Convenient Transportation"

Because they are 47 li from Badong and 10 li from the town of Guandukou, Yangjiapeng's villagers must rely on river transportation to maintain contact with the outside world. Early every morning, people travel to town by river carrying a basket of goods, usually fruits and vegetables. After selling their wares, they buy whatever they need and return home. During the orange-growing season, they sell oranges to vendors but also hire boats to ship their goods to Yichang, Shashi, and Wuhan. Villagers no longer have to survive on sweet potatoes and millet, they can now afford to buy rice. The staples of ten years ago are now fodder for the pigs. The river as transportation is essential for the economic vitality of Yangjiapeng. It should come as no surprise, then, that villagers insist that if they are moved a great distance, they hope to be settled in a place with equally convenient transportation.

"Grow Oranges Not Paddy Rice"

Before 1971, in accord with Chairman Mao's policy of "planting grain everywhere," the Chinese government did not allow villagers to grow anything except grain, and as a result many people were very poor.* A strong laborer could only make a few cents per day. Millet and sweet potatoes were the staple food, while the pigs fed on grass. If villagers wanted to make enough money to buy oil and salt, they had to go fishing in secret or gather firewood to sell. More recently, however, the government's control over grain production has diminished, and since 1975 villagers have been growing oranges for export. Since 1984 orange groves have proliferated, and today oranges are the primary cash crop in Yangjiapeng. Most families now own, on average, over one hundred orange trees. Orange production has changed people's lives dramatically—they now make enough money to buy rice, coal, clothes, and also put some away to build a house. In the past, people raised pigs to provide for a New Year's feast. But now pork is part of the daily diet. Most villagers feel that without oranges they have no economy. Orange production is, indeed, so important that it is said that son-in-laws from villages high in the mountains now often move in with their mother-in-laws in the valley to enjoy the riches. If the government or anyone else asks a villager to move, their first question is always: "Can we grow oranges there?"

Villagers Preference on Resettlement Policy

For villagers living in areas where moving back from the river and resettling in nearby areas is impossible, there are sometimes other options. The

---

*The policy, promoted by Mao, was part of an attempt to make China self-sufficient in grain, though it came at great cost to other agricultural products.

possibilities might include: moving to a new location near the river, but miles away from their present home; shifting to nonagricultural employment; cash compensation from the government; or moving the elderly and children back from the river and into nearby areas, while the young move away and into nonagricultural employment.

Eighty-three percent of the people we interviewed said that they preferred the last option. The elderly treasure the land left by their ancestors and they also value their friendships with neighbors. Young people, however, are forward-looking and willing to make a go of it somewhere else. They are less nostalgic about ancestral land. This option maintains the integrity of the community and the family and also allows the young to leave without breaking with their past totally.

Why Villagers Are Unwilling to Move to Urban Areas

Dongrangkou Village, which is across the river from Badong, has hundreds of houses on the flat lands between the mountains and the river. Before the Gezhouba dam was built, the river in this area was so shallow that most ships were unable to pass without the help of trackers from the village on the sides of cliffs. When the Gezhouba dam was completed, the water rose, putting many people out of work. But the villagers adapted to the changes, and in recent years Dongrangkou has become a transportation hub. A village dock was built, and it has become a major embarkation point for people to take boats across the river to Badong. Many villagers have taken advantage of the new emphasis on transportation and have built private shipping docks or bought ships or other vehicles and started their own shipping businesses. Village businesses are thriving, too. Restaurants, hotels, and stores are all flourishing. Farmers now sell their produce in Badong. Many of them can be home for lunch, their workday finished. This kind of life is much easier than that for those living back in the mountains who have to get up at 3:00 A.M. and walk more than 30 li to haul their goods to market. But when the Three Gorges dam is built, the village will be flooded. Moreover, there is no land nearby and everyone will have to be moved far from their homes.

The Yichang branch of the Three Gorges Economic Development Office once planned to move the entire village to the Wujia port area in Yichang. Representatives from the village were invited to visit Wujia, and most felt that life in Wujia port was comfortable—it was near a busy shipping dock, and most of the farmers there were engaged in commercial activities or were hired out as seasonal workers to work in orchards, hotels, or stores. Nevertheless, 80 percent of Dongrangkou's residents were unwilling to move. Instead, they have insisted on moving back from the river and resettling their village on a nearby mountain named Leijiaping. Why are they choosing a seemingly more difficult path (moving up into nearby mountains) over moving to the prosperous Wujia port area?

Put simply, the villagers do not want to move to the city. "Our present life is not so bad," they say. "We are mostly self-sufficient in food and can even afford some luxuries like roast duck and beer when we have guests." "City life is different," they claim. "Once you step outside your door, you have to pay for everything—vegetables, coal, water, grain, and transportation. In the cities, unless you have skills, you're still a farmer." In the end, the conclusion of many is that it would take years for them to establish a new life in the cities, and if they were to move they "would have to eat bitterness for many years."

Recently, there was a gold rush on Hainan Island (off China's southeast coast) and quite a few young people from Dongrangkou Village went there to seek their fortunes. Many succeeded, earning ¥400 a day. But within three months, everyone had returned to the village because life on Hainan was simply too hard. They made a lot of money, but their expenses were high and they suffered a great deal. To the disappointment of many, they could not adapt to the new environment and could not get along with the locals. No wonder most are unwilling to move to Wujia port.

Conclusion

The issue of involuntary resettlement is not simply a matter of finding land and providing financial compensation for relocatees. It is a complex process in which people must adapt to new financial, social, environmental, and employment-related situations. Resettlement will bring about important changes in people's lives—changes which our survey suggests villagers will resist.

Chapter Seven

A Survey of Resettlement in Badong County, Hubei Province

Ding Qigang and Zheng Jiaqin, 1992

Developmental Resettlement

There are two basic approaches to population resettlement resulting from dam and reservoir construction. The first is straight compensation, in which the government provides cash and land directly to individuals as reimbursement for losses resulting from inundation. In the second, the government not only reimburses the people for their losses, but also provides for adequate housing and other basic infrastructure in the new location. Experience has shown, however, that neither approach is very successful.*

In the Three Gorges project, a new approach to resettlement, called "Developmental Resettlement" (*kaifashi yimin*) is being attempted. Rather than concentrating on the direct compensation of individuals as the other two approaches, Developmental Resettlement professes to provide economic benefits to rural resettlers through government-financed reclamation of higher-elevation land, cultivation of cash crops, and investment of limited funds into industry and agriculture to create new jobs, in addition to compensating individuals. The policy is designed to enhance the long-

---

*China currently has approximately 10.2 million officially classified "reservoir relocatees" (*shuiku yimin*). This group includes those who were forced to relocate from dam projects built since the mid-1950s and their descendants. The vast majority of them live in the countryside, and about half were uprooted for major projects financed by the central government. See, Jun Jing, "Rural Resettlement: Past Lessons for the Three Gorges Project," *The China Journal*, no. 38 (July 1997), pp. 65–92.

term capacity of the people to make a living and to facilitate resource development in their new locales.

Debate over this third approach to resettlement has been going on for years. The crux of the debate is over whether local environments can support the proposed developments. Other key issues include planning for nonagricultural employment, the ability of the people to adapt to new natural and social environments, and the redistribution of resources within resettled communities.

With these issues in mind, in July 1992 we accompanied the Beijing University Three Gorges Project Research Group (*Beida sanxia shijian kaocha tuan*) to the Three Gorges area. The group's visit lasted 15 days, during which time we conducted fieldwork at trial resettlement projects in Badong County, Hubei Province.

Badong County

Badong County is an autonomous area of the Tujia and Miao people situated in western Hubei Province.* The county is located in the central region of the Three Gorges between the Xiling and Wu gorges, about 60 kilometers west of the proposed dam site at Sandouping. Wu Mountain is directly to the west of Badong, and to the east is the town of Zigui. At its most northerly point, the county is bordered by the Yangtze River, and at its most southerly, by the Qing River. These boundaries have created a long narrow county measuring 135 kilometers by 40 kilometers, and covering 3,219 square kilometers. The county consists of 26 townships, 481 village committees (*cunmin weiyuan hui*), 3,925 village groups (*cunmin xiaozu*),** and 13,081 households, with a population of 480,000.

---

*The Miao and Tujia are a tribal people of the Sino-Tibetan language group who have often been regarded as inferior Han. They have been part of the Chinese empire since the Mongol conquest in the thirteenth century. Their populations are located primarily in China's southwest provinces, but also extend into Vietnam, Laos, and northern Thailand. According to the 1990 census, the two groups number 7.4 and 5.7 million, respectively. Autonomous regions and areas in China are granted limited self-governance, religious freedom, and greater freedom from China's strict population control policies, but this has not always protected minorities from the imposition of centrally determined policies, such as the Three Gorges dam.

**In 1987, the village committees, along with the townships, replaced the government of the People's Commune as the legal basis for "village self-government" in China. Village groups are administrative subunits of the committees.

Most of the county is mountainous, steep, and at a high altitude, with its highest peak reaching 2,300 meters above sea level. More than 70 percent of the land is both mountainous and over 800 meters above sea level. The average gradient slope of the land is 28.6, with 66 percent of it having a gradient slope of 25 or more. Only 13.25 percent of the county's land is arable, 94 percent of which is dry, nonpaddy land, and 63 percent of which is found above the 800–meter elevation mark. Badong County has no first-grade, prime arable land, and more than 75 percent of the arable land is of poor quality, below the fourth grade.*

Badong County is situated at the key hub of the western Hubei transportation network, and its transportation system is relatively well-developed as compared to adjacent areas in the western part of the province. Highways crisscross the county in both north-south and east-west directions, and the Yangtze and Qing rivers flow through the county for 39 kilometers and 37 kilometers, respectively.

The county is primarily an agricultural area with millet, wheat, sorghum, paddy rice, and sweet potato production. Vegetable oil products include rapeseed, sesame, and peanuts. Cash crops include oranges, peaches, apples, and peas. In 1991, the total value of agricultural production was ¥290 million. Important local industries are mining, hydropower, fertilizer production, agricultural machinery, printing, brewing, construction materials, and food processing. The county boasts 79 enterprises, employing more than 7,500 workers. In 1990, its fixed assets were ¥100 million, and in 1991, the total value of industrial production was ¥150 million.

The vast majority of Badong's residents are adults, and the median age is twenty-six years. Most have only completed elementary school, and a labor surplus currently exists.

Resettlement Planning and Trial Projects in Badong County

Based on 1991 data, the Three Gorges dam is expected to submerge 11 townships and affect 220 village committees and 70 village hamlets in Badong County. It will directly affect 13,337 people, 39 percent of the

---

*Any land ranked below the fifth grade cannot be used productively for agriculture, while sixth-grade land will only support forest cover. Seventh- and eighth-grade lands are deserts.

total number to be resettled in the county, who will have no choice but to move.* In addition, 5,954 mu of arable land, 6,313 mu of orchards, and 93 kilometers of roads will be submerged.

In 1985, a resettlement bureau was established in Badong County. Following the central government's policy to "develop resources, move people back from the river, settle them in nearby areas, and focus on macroagriculture," the county coordinated resettlement programs for townships, villages, and village groups. To determine whether conditions in a particular area would allow people to move back from the river and settle in nearby areas, the county undertook its planning from the bottom up, that is, from village groups, to village committees, to townships, and then to the county. The county then selected areas where the policy was likely to be successful, later expanding the trial projects to the rest of the county.

Of the 13,337 people to be resettled, the county plans to place about 8,000 in agricultural employment, 4,000 in light and service industries, and 1,000 in the tourist industry. Five of the county's districts (*qu*) have been chosen to receive most of the relocatees: In the two valley areas, oranges and other seasonal fruits will be grown; in the Shennongxi district, orchards, tea gardens, and tourism will be developed;** in Guandukou subdistrict, oranges will be grown and river ports built; and in the suburbs of Fuling Township, vegetable production and industry will be developed.

Over the past six years, Badong County has invested ¥40 million for trial resettlement projects in eight townships. In total, 6,600 mu of mountainous land has been developed at 90 different sites. The county has established five trial agricultural projects to conduct research and to provide technical training for the locals. These township-level projects have developed 1,228 mu of land, 353 mu of which will be used to settle intervillage relocatees.

An additional 4,900 mu of newly developed land (at 79 different sites) has been earmarked for intravillage group resettlement. On this land, over 3,000 mu of orange groves have been planted, the first batch of which has just begun to produce fruit. In addition, 27 water storage and irrigation facilities have been built with a capacity of 14,000 cubic meters. At the

---

*Twenty-one thousand others whose land will not be submerged by the reservoir are also slated for resettlement for a variety of reasons related to the project.

**There is evidence that opium poppies too are being cultivated in the higher elevations of Badong County. See Epilogue.

same time, the county has also encouraged more farmers to grow oranges and other drought resistant crops.

Overall, these trial projects have been successful. Data from five of the trial projects reveal that total incomes and per-mu incomes have increased dramatically—from ¥58,309 with per-mu yields of ¥64 in 1987 to ¥174,990 with per-mu yields of ¥193 in 1989.

There are, however, a number of problems with the trial projects. Land available for resettlement is limited, and this does not take into consideration the fact that people may be forced to resettle for a second or third time when the original resettlement sites become uninhabitable or untillable. There is also a shortage of resettlement funds, with most of the money being targeted primarily for individual compensation. At the same time, newly developed lands lack investment for irrigation systems and fertilizer production; there has been inadequate planning for placing people in industry and tourism jobs; there is a lack of technically trained personnel and the population is poorly educated; the system for redistributing and managing land and other resources is poorly developed; and, clan forces have an inordinate influence on village government—without their support, village administrations have been unable to take action on trial projects. In short, despite the lofty claims of Developmental Resettlement, resettlement is proceeding in the old way and encountering familiar problems.

Problems with the Trial Projects

Limited Agricultural Capacity of the Region

Dam supporters claim that there is plenty of good agricultural land for resettlement in Badong County. With 280,000 mu of undeveloped land available, and only 13,337 people to be resettled, there should be more than enough land—about 21 mu per person. But dam supporters have not considered the quality of the land or the wishes of those being moved in their calculations. One of the trial sites illustrates the problem.

This particular trial site is comprised of two mountains and one valley, with the mountains surrounding the Yangtze River on its northern and southern sides. Most of the region has a very high elevation. Its highest point is 1,722 meters above sea level, and its lowest is 66.8 meters above sea level, with an average altitude of 657.5 meters. Significantly, undeveloped land more than 600 meters above sea level is not suitable for growing oranges and, therefore, is not agriculturally useful for local residents. Moreover, according to China's Water and Soil Protection Act, land with a gradient slope greater than 25 cannot be used for agriculture in terraced fields. The average gradient slope of the land at the trial site is 27.8, and 61 percent of the land has a gradient slope greater than 25. Finally, 10 percent of the land consists mostly of rock. These facts alone mean that the total amount of land available for development is not 280,000 mu, as dam supporters claim, but rather 80,000 mu. Moreover, what good land does exist is further threatened by deforestation and consequent soil erosion, and the long-term impacts of human habitation. Altogether, these factors contribute to the fact that nonhumus-type land that is inappropriate for any cultivation whatsoever constitutes 84 percent of the [280,000 mu] of barren land.

Social and cultural factors further limit the availability of undeveloped land for resettlement. According to our research, people slated for resettlement who have been living by the river are, at most, only willing to move halfway up the mountain and are completely unwilling to move on to the other side. They argue that moving from the riverside to the high mountains will radically change their lifestyle and make it much more difficult for them to make a living. In 1966, to alleviate overpopulation, a group of farmers were relocated to a lush area 1,800 meters above sea level. That site was on a plateau and had more fertile soil than do the currently proposed sites. However, most of those moved soon returned to their riverside homes, even though this meant converting to nonagricultural work as they had forever lost their farmlands.

Altogether, these natural, social, and cultural factors further reduce the actual amount of land available for development to 15,000 mu. There are 13,337 people slated for resettlement. But it is likely that some of them will have to be resettled a second or third time when their original sites become uninhabitable or untillable. Some predict that by 1998 second and third resettlements will raise the total number of relocatees to 23,000. These facts indicate that because the land-to-person ratio is extremely tight, the environmental capacity of the resettlement site will be severely taxed by the resettlers. As resettlement begins, more difficulties will likely become evident. In fact, there have already been problems in planning the resettlement of people now involved in nonagricultural employment. All this proves what Vice Premier Tian Jiyun said during his inspection tour of the Three Gorges' reservoir area in March of this year: "[We] cannot say categorically that the geographical capacity of the reservoir area which is slated to accept hundreds of thousands of relocatees is not without its problems."

Limited Employment Opportunities in Heavy Industry and Construction

Because there simply is not enough agricultural land, Badong County will have to create a large number of jobs in the heavy industry and construction sectors. But this will be a difficult task in a county where per capita income is less than ¥300 and industrial production is less than ¥1,000 per capita.

The decision to launch the Three Gorges project was delayed for decades, and neither the central nor local governments invested much in the region during that time. As a result, Badong does not have a strong industrial base. As of 1991, the county had only ¥100 million in fixed industrial assets (including private, collective, and state assets), and agricultural output out-stripped industrial output four to one (the national ratio is the inverse, one to four). In terms of employment, 86.4 percent of Badong's workers have jobs in agriculture, and only 3.7 percent in industry. Nationally, agriculture em-ploys 60 percent of Chinese workers, and industry 22 percent.

Most of Badong's industrial products are used to meet local demand, a very small market. As a result, the enterprises are small in scale, with each employing an average of 110 workers. Capital equipment is old, and pro-duction technologies are outdated. With such a limited industrial base, the county is unlikely to develop industry quickly enough and on a sufficient scale to provide jobs for the large number of relocatees.

The county is also lacking the capital to finance further industrial de-velopment. The dam is an excellent source of funds, but it is not enough. The resettlement budget includes funds for industrial development, but the primary recipients of the funds will be individuals. Relocatees should not expect large sums from the resettlement budget to be spent on industrial or other projects.* Funds from outside the county are also likely to be very

---

*By 1991, the government began to downplay the promise of Developmental Resettlement and of moving relocatees out of agriculture and into industrial employment. Authorities admitted that the industrial enterprises that were created at the experimental stage did not provide the promised jobs to poorly educated rural resettlers. See Jun Jing, "Rural Resettlement," p. 35. Chinese leaders make it clear that it is up to the "reservoir region itself [to] explore ways to attract more investments. . . ." Statement by Qiao Shi, Xinhua, June 19, 1996.

limited. According to local officials, there is already a ¥500 million short-fall in government funding for the 46 industrial projects which have been authorized. Despite Vice Premier Tian Jiyun's proposal to make projects in the reservoir area a priority for investment, investors demand high returns on their commitments, and returns in Badong County can be expected to be quite low. Take construction, for example. Because the land is so steep, it is much more expensive to install foundations in Badong County than in other parts of the country. Increased construction costs alone can reduce or altogether eliminate anticipated profits.

The expansion of industry requires that workers be adequately trained and educated so that they can make a smooth transition from agricultural to industrial employment. But few of Badong's 480,000 residents have such an education. According to the fourth census, only 1,526 of Badong's residents have more than two years of college (well below the national average); 5.7 percent have a high school education; and 51 percent have an elementary school education. Thirty percent of the rural population is illiterate or semiliterate.

The geography of the county will also limit future industrial development. Located on both banks of the Yangtze River, the county has virtually no plains or flat lands. The rolling hills in the central part of the county have hindered the expansion of water and transportation systems crucial for industrial development. Moreover, the area is surrounded by mountains and is relatively enclosed, causing poor air circulation. As a result, air pollution is likely to become a problem, while the slower flow of the dammed Yangtze will also cause pollutants to concentrate in the reservoir rather than being flushed out to sea.

Badong's population is young, and because there are a significant number of people entering the job market every year, the labor force is expected to grow for the next ten to twenty years. Again, according to the fourth census, unemployment is increasing by 6.6 percent annually. As time passes, it will be increasingly difficult to find jobs for the thousands of resettlers.

Also significant is the fact that village relocatees who find jobs in factories are treated poorly as compared to regular workers, but cannot change their household registration from rural to urban. This provides an incentive for relocatees to abandon jobs created for them at the new sites and could cause serious problems for resettlement planning overall. For example, the expansion of Badong's cement plant created 100 jobs for village relocatees at a cost of ¥10,000 per worker. But of the original 100 workers, only a little over 50 still work at the plant. The others have left because as laborers they were underpaid, making far less than private vendors or self-employed truck drivers, for example. Laborers are not eligible for pay increases, bonuses, or subsidies. Their jobs are usually temporary, their work conditions poor, and their duties more physically demanding than for regular jobs. And because the relocatees do not enjoy urban household registry [which carries with it a grain ration], they must buy high-priced grain on the market. Township workers would never risk losing their urban household registrations by leaving their jobs, even if they were just laborers. But relocatees have no such reservations because they cannot change their registration.

These are potentially serious problems with resettlement planning for nonagricultural employment. If the economy is performing well and jobs are relatively easy to find, relocatees are unlikely to cause many problems for the government. But if the economy is performing poorly, and enterprises suffer a drop in production, relocatees will probably turn to the government for help; a prospect that concerns local officials.

Limited Opportunities in Service Industries

Badong County also plans to provide tourist and service industry jobs for 2,000 to 3,000 rural relocatees over the next six years. But it is unclear whether this will be possible.

There is only one internationally known tourist destination in all of Badong, and that is the Shennong stream (*Shennongxi*). Among the stream's many attractions are waterfalls where crystal clear water drops from sharp cliffs as if from the heavens, and centuries-old caves and narrow plank roads built along the sheer cliffs of the three gorges through which the stream runs. Tourists can sail down the calm sections of the stream on boats built by the Tujia people. However, two of the three gorges along the stream will be inundated by the dam, effectively destroying much of the tourist trade in the area.

Even if the scenic value of the Shennong stream were not destroyed by the dam, and the stream was heavily promoted as a tourist destination, the local infrastructure could not handle the influx of tourists necessary to support a host of new jobs. Traveling down the stream to the Yangtze usually takes less than two hours (and is often quite dangerous), but because of bad roads, it takes another two hours by bus to travel the short distance from the county seat to the stream itself. Only 120 tourists a day can be brought to the stream.

People working in tourism require even more training and education than those in the industrial sector. The Badong Tourism Bureau currently employs only 62 people, most of whom have a high school education and are women. It would be very difficult to place rural laborers in such positions. The only opportunities for male laborers would likely be operating the small Tujia-built boats that ply the Shennong stream. But the stream can only accommodate about 30 boats, employing about 100 people. This is far fewer than the 2,000 to 3,000 jobs hoped for. Finally, work on the boats is seasonal and would not provide full-time employment.

Issues of Land Ownership, Management, and Redistribution

Land is scarce and unevenly distributed in Badong County. In some of the areas where relocatees currently live, people have an average of 1.5 mu of land per capita, and after the development of new lands, they may have more, perhaps as much as 2.6 mu per capita (in Leijiaping Village, for instance). In other areas, however, relocatees currently have only about one-half mu of land per capita and will be unable to move back from the river and resettle in nearby areas (in Dongrangkou Village, for example). To resettle these people, the county government will have to acquire land and relocate them to a number of different villages.

To complicate matters, there is a growing sense of land "ownership" among farmers in the reservoir area. Some of those who own land that the county government wishes to acquire for resettlement are asking much higher prices than the government usually offers. Moreover, the farmers have insisted that the land only be leased to relocatees so that they retain ownership. Other land owners have insisted that they be given the exclusive right to develop their land, in effect barring redistribution to the relocatees. If the government and the land-owning farmers are not able to come to an agreement soon, the ability of resettlers to make a living after inundation will be seriously threatened.

For those resettlers who can move back from the river and resettle in nearby areas, the issue of land ownership is less important than how land is distributed and managed. In Badong County, three different approaches to the distribution and management of newly developed lands have been adopted. In the first, new land is distributed to the local village groups according to population and number of households, and then, following inundation, all of the land held by the village or village group [the old and new land] is divided amongst individual households. In the second approach, the entire village collectively manages the newly developed land and then redistributes it following inundation. Under the third approach, newly developed land is distributed to relocatees before inundation according to the official criteria for reimbursement. If there is any left over newly developed land, it is evenly distributed among the village groups.

Each of these approaches has its merits and shortcomings. The advantage of the first is that prior to inundation everyone in the respective village groups can enjoy the fruits of the newly developed land and that following inundation everyone shares the benefits and profits from both the old and new lands. Possible drawbacks occur mostly before inundation and include lax management of the newly developed lands by the village group leader and the likelihood that individual farmers will be unwilling to invest in and care for lands that they might have to give up after inundation. This would be devastating for orange groves which take a long time to mature and need constant attention.

The advantage of the second approach is that the newly developed lands are in the hands of the village as a whole, which encourages collective economic action or investment. In the wake of the Agricultural Responsibility System [instituted in 1978], however, there is little collective capital in the village governments, meaning that substantial village-based investments are unlikely. Moreover, in terms of land management, there is still a public rice bowl mentality—everybody counts on someone else to do the job—which would seriously reduce the likelihood of generating a profit under the collective management system.

The advantage of the third approach is that relocatee ownership rights over land are clearly designated. This option also provides incentives for the relocatees to invest in the newly developed land and to cultivate newly planted orange groves. Among the three approaches, this last one is most beneficial over the long term for land development, and it was adopted by a trial project in Leijiaping Village. However, it too has certain drawbacks: Since relocatees will, in the years prior to inundation, have claims to two plots of land (the land they received under the current responsibility system and the newly developed land), they will enjoy a substantial increase in their income as compared with nonrelocatees. Then, following inundation, the relocatees will suffer a dramatic drop in their incomes as a result of the inundation of their original plots of land. At either stage, internal strife may emerge within the communities. This was the case in the B trial project in Leijiaping, where conflicts led to the destruction of water storage and irrigation facilities and newly planted orange groves by displaced farmers.

Developmental Resettlement should both guarantee that the living standards of the relocatees will not suffer after inundation, and facilitate a smooth transfer of land ownership. The experience with trial projects in Badong County, however, indicates that it is very difficult to achieve these goals concurrently.

People's Attitudes in the Proposed Reservoir Area of Badong

As part of our fieldwork, we interviewed a number of county leaders, the majority of whom saw resettlement in a positive light. They claimed that the Three Gorges project will stimulate the economy of the reservoir area, provide economic opportunities for its residents, and help local communities to establish ties with the outside world. To this end, Badong County has come up with slogans like "Take advantage of the Three Gorges project to create a prosperous Badong economy." The officials believe that years of indecision about the Three Gorges held back government investment in the county, and they therefore hope that the dam will bring with it an enormous influx of investment.

In our view, there is some truth to these views. To a certain extent, the project will provide a boost to the economy of the Three Gorges area. However, for external investment to have a real impact on the county's development, considerable effort is required. First, we must disabuse ourselves of the notion that external investment ensures internal development and that the government will take care of everything. Instead, development plans must be linked to local economic conditions. Second, government investments must be used to develop the local resource base and support the local economy and ensure reinvestment in it, rather than being siphoned off through corruption and embezzlement. Third, we must realize that the project will not bring wealth overnight. In fact, since government funding is limited, there will be a shortage of capital funds in the area for some time. Leaders in the reservoir area must take full advantage of special economic policies to make up for the lack of funds.* Fourth, government investments must be linked to the development of local human resources, since human resources are key to the creation of a productive economy and higher incomes. Investments in education and personnel training may not solve any immediate problems, but they are the key to future progress.

The opinions of local relocatees are, by contrast, quite complex. Experience from the trial projects indicates that people whose villages can be resettled nearby generally favor resettlement. These people still feel, however, that without the promised benefits of Developmental Resettlement, their standards of living will drop precipitously after the dam is built. People whose villages cannot be resettled nearby feel even more strongly that their standards of living will suffer as a result of moving. They worry that they will lose contact with relatives and friends and that they will have no one to turn to in times of trouble after the move. Most of those working in the industrial sector will agree to move, but they worry that their businesses will suffer and their incomes will drop. Finally, farmers living in the mountain regions have always led a hard life, and many seem willing to give up their rural lives for jobs in the city,** but fishermen and farmers living on the banks of the Yangtze enjoy a good life and do not want to work in factories or move to other towns.

Conclusion

The policies of "Developmental Resettlement," "moving back from the river and resettling in nearby areas," and "giving priority to macro-agricultural development" only provide the possibility of solving the various problems associated with resettlement. The policies' ultimate success depends not on the people's will, but on the recognition and proper management of the ways in which environmental, social, economic, societal, and human factors affect the policy goals. The discussions in this article have been limited to the economic and social dimensions of the policies and suggest that sound investments in and the maintenance of the reservoir's environmental capacity are key to resettlement success.

---

*These include open-city, custom-free status for imported technical equipment given to the ports of Chongqing, Fuling, Wanxian, and Yichang, and the creation of the Three Gorges Economic Development Zone.

**This seems to be the case of the Tujia minority, who believe the dam will improve their currently miserable economic conditions.

Though we did not focus on them nearly as much, environmental and ecological concerns are equally important. To date, Developmental Resettlement has only been attempted in small trial projects in areas that are well suited for development. But inundation has not yet occurred. As the project progresses, many new problems are likely to surface, and both present concerns and unknown future ones will require serious study.*

*Central government officials agree that "despite certain achievements, much is left to be done in resettling residents affected by the project." Statement by Qiao Shi, Xinhua, June 19, 1996.

Chapter Eight

Resettlement in the Xin'an River Power Station Project

Mou Mo and Cai Wenmei

The Xin'an River power station (*Xin'anjiang dianzhen*) was the first large-scale hydroelectric project designed and built in mainland China with domestically produced equipment. The Xin'an River originates in the Yellow Mountains in Xiuning and Qimen counties, Anhui Province, flows through Tunxi and Weng counties in Anhui and Chun'an County in Zhejiang, joins with the Lan River at a point southeast of Mei Township, Jiande County, in Zhejiang, and then flows northeast into the Fuchun River. It is 261 kilometers long and drains an area 11,800 square kilometers in size. The river is full of twists and turns and can be quite turbulent at points. Between Tunxi and Tongguan Gorge in Jiande County, a distance of 170 kilometers, the Xin'an drops a full 100 meters. For years, scientists and engineers had dreamed of building a power station on the Xin'an, but the dream was not realized until after the Communist revolution in 1949.

The Xin'an power station was first planned in 1952. In 1956, the initial design was completed and approved by the State Council. Construction began in 1957 and was completed three years later, when the first group of turbines began to generate power.

The Xin'an power station has an installed capacity of 662,500 kilowatts and generates an average of 1.86 billion kilowatt hours of electricity per year. Since the station's completion in 1960, Xin'an has become an important part of the electrical grid and has provided eastern China with a reliable supply of power. By 1984, the dam's total electrical output was valued at ¥2.27 billion—about five times the total cost of the project. The dam's massive 580–square-kilometer reservoir (the Thousand Island Lake [*Qiandaohu*]) can store 17.8 billion cubic meters of water and has provided flood relief for the people living on the banks of the river and for

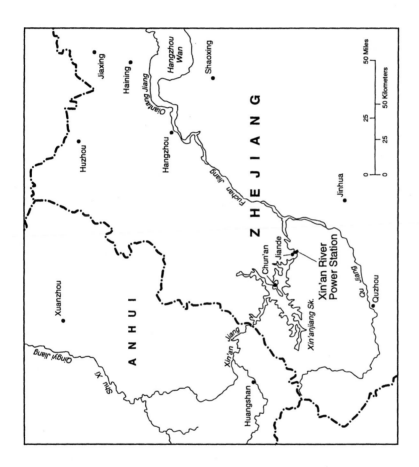

Xin'an River Power Station (Anhui and Zhejiang provinces).

hundreds of thousands of mu of farmland in the lower reaches of the river. The dam has provided substantial benefits for the people of Jiande, Tonglu, and Fuyang counties.

Above the reservoir, the main stream of the Xin'an is 100 kilometers long, and its waters are deep enough to accommodate several-hundred-ton ships. Below the reservoir, to the Fuchun River power station, 100– to 200–ton ships can now ply waters which, before the dam's construction, were virtually nonnavigable. Fisheries and the tourist industry have also been developed in the reservoir area, and the climate has changed, making it possible to grow fruit trees and expand local forests.

But the project has had some serious adverse effects too. Two counties with very long and rich histories—Chun'an and Sui'an [sic, Lin'an]—were inundated by the reservoir. Seven townships, 1,140 villages, and 320,000 mu of arable land were submerged and 280,000 people were resettled. Centuries-old cultural antiquities, factories, and over 500 middle and elementary schools were also submerged along with some of the most fertile land in the area. (The value of the property lost by the 280,000 relocatees is beyond calculation, and many relocatees have yet to be fully compensated.) The effects on local agriculture were disastrous: Chun'an went from producing a grain surplus before the dam's construction, to importing over two million tons of grain in one year after the dam was built.* The reservoir also flooded four major roads and destroyed the area's transportation system. The navigation route along the Xin'an that led directly to Hangzhou was cut off by the large reservoir, causing a major increase in the price of goods and labor in the city. The light industry sector also suffered. By 1963, the total value of the sector's output was down 86.7 percent from what it had been in 1957. Time has been slow to right the situation: By 1982, light industrial output was still 25 percent below the 1957 level, making Chun'an the poorest county of the seven that are part of the administrative area of Hangzhou Municipality. One of the many reasons for the county's slow recovery from the effects of the dam's construction is the fact that teachers, artists, and technical workers were forcibly resettled out of the area when the dam was built.

---

*In 1957, before the dam was built, 74,000 households produced 13.6 million kilograms of grain which was delivered to the state under the planned purchase and supply system. But in 1982, following the inauguration of agricultural reforms in 1978, even after a bumper harvest, 63,000 households lacked sufficient food and so the government had to provide them with 2.5 million kilograms of grain.

Xu Zhishi and the Decision to Dam the Xin'an

During the War of Resistance against Japan (1937–45), the man who would become the chief engineer of the Xin'an power station, Xu Zhishi, designed and helped build a power station near the city of Changshou in Sichuan Province. Following the Japanese surrender, Xu traveled with some of his coworkers to the city of Nanjing and then to Zhejiang Province where he coordinated studies of the Qiantang, Xin'an, and other rivers. In May 1945, three days after the city of Hangzhou was liberated by the People's Liberation Army (PLA), General Tan Zhenlin met with Xu and a group of his colleagues, complimented them on their great accomplishments in hydroelectric construction in Zhejiang, and voiced his support for similar projects in and around Hangzhou. Xu and his coworkers were excited to have a PLA commander so interested in their work so soon after the army had crossed the Yangtze. In 1952, after finishing work on another power station (the Huangtankou at Quzhou, Zhejiang Province), they started planning the Xin'an River power station. Xu felt that there were two possible designs for the project. The first, and relatively more attractive design, was a series of small, multi-tier power stations along the river. This design would prevent the inundation of large swaths of land and avoid the resettlement of large numbers of people, though it would submerge the major town of Tunxi, located in southern Anhui. The second design would see the construction of a single large dam and a vast reservoir.

After consulting with the affected provinces, the provincial Communist Party committees of Zhejiang and Anhui indicated that they would respect the will of the central leadership, and the decision to build a single large dam was made. Liu Lanpo, then the minister of fuel industry (*Ranliao gongye bu*), was put in charge of the project, and throughout its construction he lived at the site helping the scientists and engineers study and design the dam, thereby gaining their deep respect. The Ministry of Fuel Industry also favored the single dam design, though it realized that support for the project among the local population was a key issue.

For two years, Xu Zhishi (with help from the students and faculty of the Zhejiang Finance and Economics Institute [*Zhejiang caijing xueyuan*]) conducted exhaustive surveys of the local population, arable land, housing, factories, mines, transportation networks, communications, local fauna and flora, and cultural sites and artifacts in and around the proposed reservoir area. Xu and his colleagues estimated that 300,000 mu of arable land would be submerged by the dam and more than 200,000 people would have to be resettled. Since the population density of Zhejiang was high and the province had little undeveloped arable land, Xu concluded that the 200,000 relocatees would have to be resettled outside the province; otherwise their standard of living would suffer dramatically. Xu then undertook studies in the nearby provinces of Anhui, Jiangxi, and Jiangsu to determine whether the relocatees could be successfully resettled there.

Xu's work revealed that there were vast tracts of deserted arable land in Xuancheng, Ningguo, and Xi counties in southeastern Anhui. The land had been deserted since the end of the conflict between Taiping rebels and the Qing dynasty army in the mid-nineteenth century. Also, in nearby Jiangxi Province, there were 46 million mu of arable land which, at that time, supported only 20 million people. Anhui and Jiangxi provinces each agreed to take 50,000 relocatees and to build houses and other facilities for them. Jiangsu Province, however, refused to take anyone since authorities there believed that people from the mountains in Zhejiang would not adjust well to life by the sea. And so, Xu's original plan called for 100,000 people to be resettled outside Zhejiang in other provinces, with the remaining 100,000 staying in and around Zhejiang Province, mostly in nearby counties in Anhui and in Jinhua Prefecture in Zhejiang.

But the plan quickly ran into opposition. The Zhejiang Provincial Party Committee argued that the resettlers should not be moved out of the province because they were Zhejiang's "greatest resource." Since the central government believed that resettlement work was, ultimately, the responsibility of local governments, the authorities capitulated and agreed to allow all of the relocatees to be resettled in Zhejiang.

At that time, just after the establishment of the People's Republic of China, there were very few hydropower experts in China, and most decisions were made by the central leadership. Xu Zhishi's suggestions, made on the basis of years of experience and numerous in-depth studies, were not only rejected by the central government, but caused him to be personally criticized by the authorities as well. In the end, Xu had no choice but to follow orders and design a plan that would allow all 200,000 relocatees to be resettled in Zhejiang Province.

Resettlement for the Xin'an power station took place in three stages. The first was a trial project which lasted from 1956 to 1957, and saw 20,000 people moved. The second phase took place during the "Great Leap Forward," from 1958 to 1960, and moved 170,000 people. The third phase, called "readjustment," began in 1961 and involved the resettlement of 80,000 people (some for a second and third time). After 1974, the resettlement program was halted.

The Trial Phase

For the trial phase, two villages (Xiao in Chun'an County and Hukou in Lin'an County), and two townships (Tongguan and Chayuan), were selected for resettlement. Because property was collectively owned at the time, resettlement did not require the transfer of ownership rights.* The Ministry of Fuel Industry and local governments allocated ¥508 to each relocatee to cover the cost of shipping their belongings and the construction of a new house, and then forced them to move. Assets that could not be moved, or which relocatees found inconvenient to move, were also compensated for. Though the sums were small, most felt that the resettlement plan was fair.

During this first stage, the local governments took resettlement very seriously and approached the task with great care. Local officials explained the importance of the project to the people, which seemed to increase their understanding and support for the dam, and, in turn, led to a fairly high level of cooperation. Officials then traveled to the villages that were to receive the relocatees (primarily Fuyang and Tonglu) to talk with the locals about how many households and residents were to be resettled there. Each of these villages was to receive roughly two dozen people. The new homes were to have a considerable amount of land, and relocatees were to be provided with houses of similar design and quality as their previous ones. Cadres were respectful of both relocatees and the current residents in the areas designated for resettlement.

Many of the first people who were moved began to work in the agricultural and industrial sectors. Others, however, were put to work on the dam project itself. Resettlers from Baisha Town in Jiande County, for instance, helped clear waste from the soon-to-be inundated reservoir. What could not be carted away was burned or buried.

Overall, the trial projects of 1956–57 were successful. The relocatees

---

*Since the dismantling of the socialist People's Communes in the early 1980s, farmers have been able to lease land, and they have developed a strong sense of ownership rights which in the case of the Three Gorges project now makes wholesale resettlement of large numbers of people more difficult. See Chapter Six.

were content, they got along with their new neighbors and with local officials, and they were employed. But those who were moved later would not be so fortunate.

Problems Emerge: Resettlement from 1958 to 1960

From 1953 to 1958, during the period of the First Five-Year Economic Plan, China followed the Soviet concept of economic development based on centralized economic planning, rapid industrialization, and the collectivization of agriculture. In 1958, the country struck a dramatically new path with the Great Leap Forward, which deviated from the Soviet model and sought modernization based on mass mobilization, decentralization of authority to the provinces and below, and a policy of economic self-reliance.* Virtually everything in China was done quickly, all under the slogan "more, faster, better, and more economically." Countrywide, the people were obliged to carry out every order and instruction of the government without reservation or hesitation. They were ordered to eliminate traditions and break with past ideologies that had sustained the Soviet-style system, and to ignore the views of authoritative celebrities, especially those intellectuals and scientists who often opposed the new policy direction. In industry and agriculture, the Great Leap Forward introduced such innovations as "backyard furnaces" and the People's Communes, which were designed to achieve massive increases in production. Unfortunately, however, things did not turn out quite as top Party leaders had hoped. During this time, scientific principles were systematically ignored and, instead, superstition and boasting were the rule and foolhardy instructions [from the top leaders] were encouraged. In agriculture, people competed against one another to launch "highly productive Sputnik communes" that boasted of massive increases in production through such dubious devices as the double-blade plow and the close planting of rice seedlings. Officially, the communes were exceptionally productive, but the figures were false, nothing but deceit and lies. Nevertheless, people believed that the grain surplus was so large that only one-third of all arable land needed to be devoted to grain production; the other two-thirds could be forested or even lay fallow. People were so caught up in the campaign that some

---

*See, Carl Riskin, *China's Political Economy: The Quest for Development Since 1949* (Oxford: Oxford University Press, 1987).

claimed "an entire family could live off 0.3 mu of land" and public canteens sprouted up on the People's Communes and in urban work units offering the people as much food as they wanted for free. An irrational optimism about the resettlement of dam oustees became part of the frenzy, as the slogan "boost our energy, launch Sputniks, and speed up resettlement" indicates.

In 1958, the Jiande Prefecture Party Committee called for resettlement to be speeded up and demanded that an additional 53,000 people be resettled from Chun'an. (Originally, 24,000 people were to be resettled from there.) In total, 78,000 people were resettled to 14 counties in Zheijang including Jiande, Tonglu, Fuyang, Lin'an, Deqing, Kaihua, Lanxi, Jinhua, Quzhou, Wuyi, Suichang, and others. Another 60,000 people stayed in Chun'an but were resettled up into the hills near their current homes, or into the mountains in nearby regions. During 1958, then, 137,000 Chun'an residents were resettled. None of these people was told before they were moved where they were being sent or what the conditions were like there. Instead, they were forced to move with little or no notice as the drums rolled, bidding them farewell. The lessons of the 1956–57 trial projects were cast aside to the detriment of hundreds of thousands.

Given the speed with which resettlement work was taking place, settlers had little time to gather their possessions. Unconcerned, the Jiande Prefecture Party Committee instructed people to "bring along more good ideology and less old furniture." Resettlers were also told that resettlement should be abruptly accelerated like a "battle action" and that they were to create "combat-ready military style organizations" and adopt a "collective lifestyle," taking along only a hoe and a quilt as they made their journey by foot.

Transportation is critical to successful resettlement. During this phase, relocatees who were being moved to other counties were given a cart with which to haul their belongings, while those moving within their home county had to travel on foot and could bring only what they could carry. Firewood, grain, and cooking oil could be carried but domesticated animals could not and had to be slaughtered. For many, the journey was so arduous that they discarded most of their possessions along the way. Among the peasants in Chun'an it was customary to store grain seeds in heavy pots. But when the time came to move, the officials in charge of resettlement would not reimburse the peasants for the full value of the pots, which were too heavy to transport. The peasants got so angry that they smashed the pots and declared that the rubble should be "devoted to the construction of the dam." But most significantly, during this phase the policy for housing compensation was changed. Following class lines, poor and lower-middle class peasants were compensated, while those labeled higher class were not afforded such luxuries.

By this point, the project was proceeding so quickly that relocatees working in the reservoir area were unable to clear all of the waste out of the way of the rising water, and laborers were being sent from Hangzhou to dismantle the relocatees' houses so that they would get out as soon as possible. Normally, resettlement should precede reservoir construction, but during the "Great Leap Forward" phase of the Xin'an project, the sequence was reversed—relocatees were literally chased out of their homes by the rising water. During the move, many of the relocatees were forced to camp outside without adequate food or shelter. Some became ill, some died, and pregnant women were forced to give birth during their journey. According to Chun'an resettlement officials, the relocatees were, in effect, being exiled. And as one relocatee commented, the situation was no different from the flight from the Japanese he had seen depicted in popular war films. During 1958 alone, 59,000 relocatees from the reservoir area in Chun'an County were resettled to other production teams within the same county.*

The haphazard resettlement process caused considerable anger, conflict, and resistance. There were arguments with local officials, smashed pots, and refusals to move. Some even set fire to their homes. But all the orders were implemented, by coercion where necessary. The hearts of the relocatees were broken; they realized they were not the masters of the country.

Privately, not all officials supported the resettlement program. There were local officials and Party members among the peasantry who disagreed with the policy and saw it as a violation of the masses' interests. But these officials had no choice but to obey the orders of the top authorities. Since the 1957 Anti-Rightist Struggle [directed against intellectuals and scientists] was in full swing, all independent opinions and criticisms were suppressed and people were generally afraid to speak out. Anyone who dared voice an opinion was labeled as "anti-resettlement" and "anti-

*Production teams were the lowest level of organization in China's pre-1978 socialist agriculture, and usually consisted of several families.

Great Leap Forward" and was "subject to criticism."* Local officials were powerless to change the situation; the best they could do was help the peasants move their possessions.

As we noted earlier, relocatees were to receive ¥508 each. But that figure was later reduced to ¥478, and by the "Great Leap Forward" phase of the project it was lowered further to ¥150. In the end, most relocatees from Chun'an County received a mere ¥120.

Most of the compensation funds were earmarked for new home construction. But the houses the government built were too small and of poor quality, and many peasant families refused to move into them. The houses were only 9 by 12 meters in size and their walls were made of bamboo sheets covered with yellow mud. Because so many families refused to move into the new houses, the state lost a great deal of money. Qu County, for instance, lost ¥300,000. In the areas where "combat-ready military-style organizations" and "collective lifestyles" were promoted, housing was even worse. Again, many refused to move in, and the county had to convert the property to factory workshops at a loss.

Before the advent of the Great Leap Forward, resettlement of Zhejiang residents was supposed to be of four types: 65 percent of the relocatees would be resettled on barren, undeveloped land; 17 percent would be resettled to existing communes; 10 percent would be moved up the mountains and resettled in nearby areas; and 8 percent would be moved to newly built towns. But between 1958 and 1960, this plan was completely ignored.

Zhejiang Province already had more people than land and very little undeveloped, barren land in reserve. As a result, developing barren land for resettlement was not a viable option. Moreover, the lack of water and irrigation facilities meant that peasants who were moved to the area found it very difficult to develop the land. For instance, Shishandi Brigade** of Baima Commune had 61 households with a total population of 232 people and 159 mu of arable land, or 0.68 mu per capita. Fifty households comprising 171 people were then resettled in the brigade, reducing the amount of land per capita to 0.39 mu. Before receiving any relocatees, Daimaoling Brigade of Fuxi Commune had 59 households, a population of 250, and

*In China at this time, "subject to criticism" could mean execution.
**The brigade was the second tier of China's socialist agricultural system and usually consisted of a single natural village.

132 mu of arable land, or 0.53 mu per capita. Forty-seven households totaling 228 people were resettled there, thereby reducing the amount of land per capita to 0.28 mu. In the end, many relocatees simply moved elsewhere, exacerbating population problems in those areas.

The conditions for moving people up the mountains and into nearby areas were no better. A relocatee named Ye Jinggen moved up Chun'an Mountain with his parents in April 1960. Only 0.28 mu of land was allotted per capita. Before resettlement, most relocatees, including Ye's family, had lived along narrow strips of land next to the Xin'an River, but now they were being given pitifully small plots of land in the middle of the mountains—land that dried up quickly when there was no rain, or washed away during storms. The resettlement was being carried out without first determining how many people the mountain sites could support. Recognizing the problem, county authorities petitioned the provincial government to allow people to be moved somewhere else, but they were turned down. Instead, the provincial authorities called on people to overcome adversity and develop the land. The relocatees were understandably frustrated. And, in certain areas where schistosomiasis was rife, such as in Kaihua County, relocatees were very concerned, which made it even harder for them to settle down.

Despite all of these problems, 170,000 people were moved from the reservoir area between 1958 and 1960. The dam might have succeeded in controlling floods and generating electricity, but the problems with resettlement should not be overlooked. During this time, when the central leadership emphasized "planting grain everywhere," resettlement was, in effect, nothing more than a means of spurring more grain production. After three years of the Great Leap Forward, the country was suffering enormous losses. As a result, the disastrous resettlement policy was made even more salient.

"Readjustment": Some Relocatees Return to Their Old Villages

The second phase of resettlement (1958–60) was so disastrous that major readjustments in policy were absolutely necessary. With the start of the Cultural Revolution (1966–76), more than 10,000 relocatees moved back to Tonglu, Kaihua, and Jiande counties in Zhejiang. Those relocatees who had managed to stay in their respective counties but were unable to make a living after resettlement launched "rebellions," not to seize power but to demand that they be returned to their original way of life. Others lashed out violently against the cadres who had been in charge of resettlement during the dam's construction. A cadre named Zheng Baoxi of Zhenjia Commune was permanently maimed by angry relocatees. Another, Fang Wenlong, of the former Fangzhai Commune, was criticized for ignoring the plight of the masses and fearing the higher authorities. Fang was publicly humiliated in the village and, in the end, suffered a breakdown. Finally, the county magistrate of Kaihua County, who had embezzled resettlement funds, was captured by angry resettlers who threatened to drown him in the reservoir. Only the intervention of local army officials saved his life.*

The situation was frustrating for the regime. In 1968, twelve years after resettlement began, Zhejiang officials reconsidered their earlier decision to resettle everyone within the province and reached an agreement with Jiangxi to move relocatees there. As a result, 54,000 people were moved to Fuzhou, Jiujiang, and Ji'an prefectures. The resettlers brought with them the farming skills of the Zhejiang peasantry and produced bumper harvests. Many became model workers and most were well liked by the Jiangxi locals. For some relocatees, living conditions were improving.

Regardless of whether relocatees returned home, or resettled inside or outside the province, many moved more than twice. Statistics obtained by Chun'an County reveal that 80,000 relocatees were resettled at least twice between 1961 and 1970 at a cost to the government of ¥40 million. And, according to a study by the Ministry of Water Resources and Electric Power in the early 1970s, 86,000 relocatees who were resettled a second time were moved outside their home provinces after their first resettlement sites became uninhabitable or the land untillable. If we include 19,000 relocatees who moved to Jiangxi and Anhui voluntarily, the total number of relocatees resettled outside their native province constituted 40 percent of the total number of relocatees. Experience with resettlement proved that Xu Zhishi's original proposal that half of the relocatees be resettled in Zhejiang and the other half be moved to nearby provinces was a sound approach to the problem.

---

*Li Rui, a vice minister in the Ministry of Water Resources and Electric Power in the 1950s, reports that one top official in charge of resettlement work in China was murdered by angry petitioners during the Cultural Revolution. See, Dai Qing, *Yangtze! Yangtze!*, pp. 130–31.

After the dam's completion, mismanagement of the water levels in the reservoir led to further problems with resettlement. The water level was persistently too low, which encouraged people to refuse to move out of the area even though they were in danger of being flooded. It also encouraged those who had already moved, but were unhappy with their resettlement, to return to villages in the reservoir area. At that time, the State Council-set normal pool level for all large hydroelectric dams was 108 meters. However, according to documents provided by Chun'an County, between 1961 and 1980, the average water level at Xin'an was below 100 meters 93 percent of the time. During that same period, the average pool level was 90.1 meters.

Because the pool level was so low most of the time, significant amounts of land below the 108–meter level were left exposed. Between 1961 and 1965, the low water level lured about 20,000 homesick relocatees back to villages and towns in the 100-to-108–meter level. They were met, however, by government officials who insisted that they move again. Some refused to leave, while others complied only to turn around and try to return to the reservoir later. The relocatees were willing to accept the risk of being flooded off the land because every additional harvest they were able to eke out brought in much needed income. To make matters worse, public facilities, including permanent structures like factories, warehouses, docks, roads, railway lines, and train stations, were also being built below the 108–meter normal pool level. Meanwhile, some technicians and engineers were questioning the scientific basis of the 108–meter normal pool level. These factors strengthened the people's resolve to stay in the reservoir area, and made it increasingly difficult to resettle the roughly 60,000 people still living there; the people had strong arguments for refusing to move,* their intransigence is what probably led the State Council to lower its standard 108–meter normal pool level [for large dams] to 103 meters in 1966.

Then, in June 1973, the waters at the Xin'an dam rose to over 106 meters. People were forced to flee the area, and nearby peasants lost their crops and other enterprises. The construction which had taken place below the 108–meter mark had proceeded despite the availability of resettlement funds to move people out of the area. The State Council had provided funds for relocatees living below the 105–meter water level, and, in January 1974, it granted an additional ¥300 million for relocatees living be-

---

*This phenomenon is also already evident in the Three Gorges project area.

tween the 105–and-108–meter levels. Zhejiang Province alone was granted ¥22.8 million for resettlers in the 105– and 108–meter zone, but it used the money for those living below the 103–meter water level instead. In response, the government chose to punish those living below the 105– meter mark and, in 1974, the State Council decided to resettle more than 10,000 people from the reservoir area to Jiangxi Province. But the political turmoil surrounding the "Anti-Lin [Biao] and Anti-Confucius Campaign" [1973–75] made it impossible to carry out this decision.*

In 1978, the Zhejiang provincial government stopped all resettlement programs and allowed people to stay below the 108–meter level. It also decided not to punish people who had built below the line, and allotted peasants grain requisition targets—a formal recognition of their settlement below the 108–meter line. This formal recognition indicated that the government had lost control of the resettlement situation.

In 1980, floods caused the water level in the reservoir to rise once again, this time to over 107 meters. The reservoir performed as planned, storing flood waters and protecting people and property downstream from the dam, but the flood was devastating for people living below the 108– meter line in the reservoir area. For the second time in seven years they were flooded from their homes. Numerous villages were submerged and all of the crops were destroyed; people were forced to high ground and were surrounded by flood waters without adequate food or fresh water. They had become refugees.

By the end of 1982, ¥224 million had been spent on resettlement—almost twice the original budget of ¥113 million. Then, in 1983, disaster struck again. Floods raised the water level over the 107–meter level once again. Tens of thousands of people managed to escape but about 10,000 people were caught in the water and most of them perished. This last flood proved once again that the 108–meter normal pool level approved by the State Council had been correct.** After the disaster in 1983, the government was forced to provide ¥50 million in emergency relief money.

---

*The "Anti–Lin Biao and Anti-Confucius Campaign" saw a struggle for power among the top leadership in the waning years of Mao Zedong (who died in 1976). The campaign pitted major bureaucracies and factions against one another and led to a slow down of government operations in China and an inability to implement policies.

**Though this policy of adopting a lower normal pool level and allowing residents to remain in the area between it and the crest of the dam was a dismal failure and led to many deaths at Xin'an, it is the same policy being proposed for the Three Gorges dam. See Li Boning's comments in Chapter Four, and Chapter Six.

Resettlement funds should have been invested in the relocatees' productive capacities so that they could make a living on their own. Instead, much of the money was used for emergency flood relief and to subsidize the daily needs of relocatees. As a result, the relocatees became dependent on social welfare. Though a great deal of money was spent, resettlement did not improve. Some relocatees were known to comment: "Funding this year and subsidies the next, yet we still suffer."

By 1983, according to the Chun'an County Resettlement Office, there were 65,000 relocatees living on 20,000 mu of arable land in the county, or 0.34 mu per capita. Seventy percent of the relocatees had less than 0.35 mu of land. For most, this was less than they had before resettlement. The county's per capita income was ¥205, but 61.4 percent of the relocatees earned less than ¥200 per capita, and 28.2 percent of them earned less than ¥150. In order to make a living, many of them had to seek out work other than that to which they were accustomed, a task which they found very difficult.

Where transportation was convenient and training readily available, some farmers quickly learned to raise crops and animals in the new environment. In Chun'an County, however, transportation was generally quite poor and there were few connections overland or by water. People were scattered about on the many islands created by the reservoir and had little contact with one another. Technicians and experts were reluctant to travel to or work in the area, and university graduates would not accept jobs there under China's labor allocation system. As a result, the poor peasants were left without any means to learn the skills needed to survive in their new environment, and many of them were forced to live off state-subsidized grain. They could not afford to buy it on the open market.

For Western tourists, however, the reservoir [known as the Thousand Island Lake region] seemed exceptionally beautiful. But foreigners had no idea how the lake had been created, and they knew nothing of the relocatees' suffering.

We once interviewed a relocatee named Wu Haiquan and his family. They had been moved four times because of the dam. After their initial resettlement, they moved back to their old home in 1962. Later they moved to Jiangxi and then to Anhui, eventually settling down on one of the islands in the reservoir. For twenty-five years, 17 families had lived on this particular island cultivating 14 mu of land. There were seven or eight houses, one made out of tile and the rest of mud and grass. Neither electricity nor gas was available on the island. In 1980, the government provided the locals with formal [rural] household registration (*hukou*) and a ration of three jin of kerosene for lighting. When the kerosene ran out they had to use candles and those who couldn't afford them had to burn pine twigs. In the middle of the living room of Wu's modest house was a portrait of then Chinese Communist Party (CCP) Chairman Hua Guofeng, who had succeeded Mao Zedong.

Wu used to have four children, but two of them drowned after falling from the steep cliffs along the island's shore. The island did not have proper docks. The deaths of his children so worried Wu that he did not allow his other children to leave the island very often and, as a result, they were unable to attend school. One of Wu's children, a daughter, was eighteen years old but illiterate. Though tourists had begun to visit the island, she found it difficult to engage with the outside world.

Lessons Learned from the Xin'an Resettlement

Resettlement for dam and reservoir construction is very different from voluntary relocation on the part of the masses. Because entire areas disappear from the map and new ones emerge, it must be well planned and well organized. This type of resettlement not only affects the lives of thousands of people and their offspring, but it can also affect the economic development, social stability, and ethnic unity of an entire area. Although the problems with resettlement at the Xin'an River power station are now history, the Xin'an experience should serve as a reference point for future large-scale resettlement. Some recommendations include:

Have a Clear Understanding of What Resettlement Is
and What It Entails

By leaving their homes and lands, relocatees made a great sacrifice for the Xin'an dam. The many who will enjoy the benefits of the power generated by the dam should be grateful to the relocatees for their sublime spirit in supporting the decisions of the country.

Relocatees are average working people. They are farmers, workers, teachers, and artisans. Since they obeyed the decisions of the government and left their homes to resettle in new places, the government should guarantee them jobs and materials, while protecting their rights as workers, and respecting their choices as much as possible. In this way, people will continue to contribute to the construction of the nation, they will be able to make a living and meet the needs of their families, and they will not suffer a decline in their standard of living.

Relocatees should be reimbursed by the government for losses incurred during the resettlement process, and those seeking reimbursement should not be treated like beggars.

Balance Responsibilities

Those in charge of resettlement work should balance their responsibilities to higher authorities with an equal sense of duty to relocatees, minorities, and scientists. Operation of the reservoir and electricity generation must be balanced with planning and arranging new work for relocatees, guaranteeing new housing for them, educating them about the project, and making sure that resettlement is carried out on schedule. Fulfilling the tasks laid down by higher authorities and guaranteeing new and productive lives for the relocatees should not be mutually exclusive.

In the case of the Xin'an power station, however, this balance was not kept. There were major errors in resettlement, and many relocatees suffered greatly. Relocatees must not be made into welfare recipients. Those resettled for Xin'an felt swindled and unhappy and they have become an unstable force in society. As it now stands, whenever potential relocatees hear the words "water, electricity, and resettlement," they can hardly contain their fear. In our future work, we must pay attention to these problems and try to solve them before they arise.

The interests of local minorities must also be balanced with the tasks assigned by the higher authorities. Our country is rich in water resources but, to date, hydropower still constitutes a very small percentage of the country's total electrical output.* Most of our water resources are located in southwestern China, in areas heavily populated by minorities. For instance, the Jinsha River flows through Qinghai, Tibet, Yunnan, and Sichuan provinces. Its middle and lower reaches are located in the Tibetan Minority Autonomous Region of Diqing in Yunnan Province, the Naxi People's Autonomous County in the Li River, the Bai People's Autonomous Prefecture of Dali, the Yi Minority Autonomous Prefecture of Chuxiong, and the Yi Autonomous Prefecture of the Liang Mountains in

---

*In 1994, 19 percent of China's electricity was generated by hydropower.

Sichuan. Therefore, if the Jinsha River is developed, it must be done in cooperation with the peoples of those regions. The development of water resources can bring both material and spiritual wealth to people, but we must also respect their local customs and practices and protect their forms of production and lifestyles.

To achieve this balance in developing hydropower resources, we must also rely on thorough scientific and technical analysis. In other words, the enthusiasm for socialist construction on the part of our leaders must be integrated with the spirit of exploration, creation, and realism on the part of our scientists.*

The Xin'an River power station was a masterpiece of self-reliant design and construction by Chinese scientists and technicians. Our people were extremely happy and proud of the dam, which everyone agrees was well-built. But at the same time our understanding of the complicated nature of resettlement work was lacking.

The first national census conducted in 1953 provided the government with a rich array of data and facts. Zhejiang Province had a population density of 224 people per square kilometer which was more than twice that of Jiangxi Province (102 per square kilometer), and greater still than Anhui Province. These data clearly indicate that in Zhejiang there were more people than the land could support. After two years of exhaustive study by hydrologists and after extensive consultations with officials in Jiangxi and Anhui, all parties agreed that half of the 200,000 relocatees from the Xin'an reservoir should be moved to the two provinces. This scientifically based decision should have been respected.

But it was not. All the relocatees remained in Zhejiang, placing an enormous burden on the province. Three hundred thousand mu of arable land was submerged by the dam, without any improvements in agricultural production. The decision, in 1968, to resettle 50,000 people to Jiangxi Province alleviated the problem somewhat.

The conditions relating to resettlement in Anhui should also be recalled. Before 1949, Jingde County in southeastern Anhui had a population of only 56,000, but by 1987 the population had grown to 148,000. Anhui as a whole had grown by only 75 percent. When we were passing through southern Anhui in 1987, we happened to stay in Jingde.

---

*Political interference in scientific issues in China is documented in H. Lyman Miller, *Science and Dissent in Post-Mao China: The Politics of Knowledge* (Seattle: University of Washington Press, 1996).

Jingde County has 169,000 mu of arable land. The rice paddies yield two harvests a year and produce about 750 kilograms per mu. In 1986, per capita income in the county was ¥500 and the county had a grain surplus. In short, conditions in the county were appropriate for receiving relocatees. Indeed, if the original proposals of the scientists involved in the Xin'an project had been followed (to resettle one-half of the relocatees in Jiangxi and Anhui), the relocatees would have been of immediate use in areas such as Jingde where they could have contributed to national construction, and the government could have saved a great deal of money from its resettlement budget. All in all, the experiences in resettlement for the Xin'an River power station should serve as a lesson for political leaders on respecting the opinions and decisions of scientists.

Well-Staffed and Responsive Project and Resettlement Offices

In any large project such as Xin'an, project and resettlement offices must be established and staffed with officials who are willing to serve the people and who are able to integrate the people's needs with their assigned tasks. Developing water resources, building power stations, and resettling people are large-scale tasks. Plans must be designed and implemented, but the opinions and sentiments of the relocatees must also be understood. Their lifestyle and work must be maintained, and new avenues must be explored to increase their income. Relevant departments from the central and local governments must shoulder these tasks together in the spirit of serving the people, and ensuring that future generations share in the benefits. The offices in charge of the Xin'an project did not do a good job of fulfilling these tasks.

Site Visits, Meetings, and Monitoring

Listening to the opinions of the masses will lessen their suffering and help prevent "rebellions" by tens of thousands of people. It can also help reduce the costs of resettlement for the state. Regular meetings should be convened where representatives of the relocatees can voice their views on problems with reimbursement, and the difficulties they are having maintaining their standards of living and work. Resettlement in the Xin'an project affected 14 counties and municipalities in Zhejiang and 16 counties in Jiangxi. Knowing how that work was carried out and at what stages problems and solutions emerged are all matters that need to be discussed. This type of communication can only benefit both relocatees and resettlement officials.

Finally, it would be useful to organize units to visit the resettlement sites and learn about the lives of relocatees following their resettlement. Delegates to the various people's congresses, political consultative bodies, members of China's democratic parties* and mass organizations, and scientists and experts should be asked to visit and help monitor resettlement sites after people have been moved. Since the offices overseeing resettlement work have little real power and the accounting and auditing systems in China are unsound, it behooves us to invite these people to learn about, monitor, examine, and help in this work.** Even if resettlement offices were strong and the auditing system sound, bringing outsiders in to oversee some of the tasks involved in dam construction would encourage a democratic spirit. People from all walks of life should be willing to participate in this process.

---

*China's eight major "democratic parties" are largely powerless organizations which did, however, express opinions on policies at odds with the CCP (especially on water conservancy and reservoir construction) in the 1950s and mid-1980s. See Dai Qing, *Yangtze! Yangtze!*.

**The State Auditing Administration and the Ministry of Finance have been given the authority to monitor expenditure of these funds. Xinhua, February 2, 1996.

Chapter Nine

The Danger to Historical Relics and Cultural Antiquities In and Around the Three Gorges Area

Interviews with the Director of the National History Museum of China, Yu Weichao

Dai Qing

Author's Note

Yu Weichao is an archaeologist and expert on the culture of the Chu kingdom (700?–221 B.C.) and on the history of the Qin (221 B.C.–206 B.C.) and Han (206 B.C.–A.D. 220) dynasties. In 1961, Yu graduated from the History Department of Beijing University with a major in archaeology. He later received a master's degree from the university, where he has periodically since taught in its history and archaeology departments. Currently, Yu sits on the Board of Directors of the China Archaeological Association and the Chu Culture Association, and is director of the National History Museum of China. His archaeological digs include the Shang dynasty (1766?–1122? B.C.) site at Panlongcheng in Hubei Province, the Western Zhou dynasty (1122?–771 B.C.) site at Shaozhen in Shaanxi Province, the Neolithic (c. 15,000–1766?B.C.) Kayue culture site at Suzhi in Qinghai Province, and the Neolithic and Warring States (472–221 B.C) period site *at Zhouliangyuqiao in Hubei Province. Among his major publications in Chinese are* Historical Legacies and Remnants of Grain Water Transport in the Three Gate Gorge *(1959),* Collection of Articles on the Pre-Qin and Han Dynasties *(1985), and* An Investigation of the Commune System in Ancient China *(1988).**

First Interview, October 1994

Dai Qing: We all know that there are many archaeological sites in the region of the Three Gorges [Qutang, Wu, and Xiling gorges] along the Yangtze River. The possible inundation of their invaluable ancient relics concerns everyone. Among the 412 experts involved in the assessment of the Three Gorges dam project, there was not one sociologist, cultural anthropologist, or archaeologist—it's beyond belief! Now that construction has begun, what do you think will happen to the area's treasure trove of historical relics and cultural antiquities?

Yu Weichao: It is true that not one archaeologist was consulted during the project assessment. Earlier this year [1994], however, the Three Gorges Construction Committee (*Sanxia jianshe weiyuanhui*) and the State Bureau of Cultural Antiquities (*Guojia wenwuju*) formally designated two units to undertake preservation and protection of archaeological sites in and around the Three Gorges dam area. One unit is our history museum and the other is the China Cultural Antiquities Research Institute (*Zhongguo wenwu yanjiusuo*). We were assigned responsibility for subsurface sites, while the Cultural Antiquities Research Institute was charged with handling the aboveground sites. Preliminary planning and survey work was already begun in November 1993, and I have been chosen to be director of the work group. Twenty-eight other academic institutes have joined in the project. We now have a basic idea of what archaeological material will be submerged by the reservoir. The institutions were to work out their own proposals first and subsequently develop a comprehensive report intended to be submitted to the Three Gorges Construction Committee by June 1995.

DQ: Are you saying that in the past you were unsure about these historical sites, or that you have made new discoveries, or have become aware of new site locations?

---

*Excerpts from these interviews were published in *Orientations* (July/August 1996): 62–64, and *Archaeology* (November/December 1996): 44–45. For more on the dangers posed to archaeological sites in the Three Gorges area see Appendix C.

Stone Fish, White Crane Ridge (*Zhongguo Changjiang Sanxia*, Hong Kong, 1993, p. 198)

YW: Among the many world-renowned antiquities in the Three Gorges area are the low water calligraphy carvings (*kushuitike*) of Stone Fish, White Crane Ridge (*Baiheliang shiyu*), which date back to the mid-to-late Tang dynasty (A.D. 618–907). Another example of low water calligraphy is the Soul Stone (*Lingshi*) site below Facing Heaven Gate (*Chaotianmen*) in Chongqing, Sichuan, which is even older than the White Crane Ridge site. According to the *Complete History of the Tang Dynasty* (Quan Tang wen), this Soul Stone site dates back to the period of the Eastern Han (25 B.C.–A.D. 220) to Jin (A.D. 266–316) dynasties. Another world-renowned site is Dragon Spine Stone (*Longjishi*) near Zhang Fei Temple at Yunyang, which is also low water calligraphy carvings. The Qu Yuan Temple, which is located in the same general area [near Zigui], has been of primary popular concern, but it is actually an example of contemporary architecture—it wasn't built until the 1980s. There is also Zhang Fei Temple built during the Northern Song dynasty (960–1126) and restored late in the Qing dynasty (1644–1911).* And then there is Stone Treasure Fortress (*Shibaozhai*) at Fuling. Although its

---

*Qu Yuan (338–279 B.C.) is perhaps China's most famous ancient poet. General Zhang Fei is the hero of the Three Kingdoms period (A.D. 220–265) who in 220 at the famous Peach Garden took an oath of loyalty in the face of death.

Shibaozhai stonehouse temple. *(Photo courtesy of Elizabeth Childs-Johnson)*

architecture dates back to the Ming (1368–1644) and Qing dynasties—not as ancient as the others—the surrounding scenery is quite extraordinary, perhaps the most magnificent in the entire Three Gorges area. All of these sites will be inundated as the dam project progresses. There is another fabulous site, the ruins of Dachang Village (from the Ming-Qing period), which is the best of its kind in the gorges area. We have proposed relocating the entire village for its preservation.

DQ: What about subsurface sites?

YW: Despite our efforts over the past twenty or thirty years, we knew little about these sites until we began our studies last November. First and

Wushan "Doomed City" with nearby Neolithic complex. *(Photo by Audrey Topping)*

foremost is the famous site of the Daxi Culture (c. 5,000–3,200 B.C.), the Neolithic complex unearthed near Wushan, Sichuan that dates back about five to six thousand years. Over the past ten to twenty years, archaeologists have worked hard to determine the western border of the Daxi Culture and also the nature of the culture that lay to the west of it, but they were unsuccessful. Now, finally, we have discovered that the western border of Daxi is in Wushan. The Daxi region is quite flat, and west of the region is Qutang Gorge, which is not very long but is extremely steep, so steep that human habitation is considered nearly impossible. On the other side of Qutang Gorge is the site of another ancient culture that we have yet to name. We now have a basic understanding of the distribution of different cultures at various times during the Paleolithic (c. 2,000,000–15,000 B.C.) and Neolithic periods in the Three Gorges area.

Entrance to Qutang Gorge. *(Photo courtesy of Jim Williams)*

Archaeologists who specialize in the Paleolithic period have un-earthed more than a dozen sites in the Three Gorges area. All of the sites are from the mid and late stages of the Paleolithic. We don't have the exact dates yet, but until recently we knew very little about the sites at all. We have come to understand the demarcation between two cultures, namely, the middle and late stages of the Paleolithic (we even found a site where stone tools were made), and the farming and fishing cultures of five to six thousand years ago, as well as the cultural system of the Jiang-Han river plain. This is invaluable information. We have come to understand that the region west of Qutang Gorge in the eastern and northern part of Sichuan Province all belongs to the same cultural system (*wenhua tixi*).

Another great discovery is the so-called Ba-Shu Culture (2,000–220 B.C.). In the past, the Three Gorges area was thought to be associated primarily with the Ba Culture, but we were unclear about where the center of the culture lay. Since the late 1970s, more digs have been carried out in the Xiling Gorge area in Hubei Province, some at very small sites, but we learned very little. (Unfortunately, we know very little about areas upstream west of Wu Gorge.) But now, finally, we have found two sites that we think formed the center of the Ba Culture.

One is Shuangyantang, on the banks of the upper reaches of the Da-ning River near Wushan. This find probably represents an early stage of the Ba Culture and is over 10,000 square meters in size. The strange thing is that it is located near the riverbed where the water level fluctu-ates a great deal. In the past, we never would have thought that such a site would lie so near a river. About five or six kilometers from Shuangyantang we discovered a very large *zun* [wine vessel] roughly 80 centimeters in height and exactly like another excavated at Sanxingdui in western Sichuan Province. We can now identify Shuangyantang as an early center of Ba Culture.

A second site, Lijiaba at Yunyang, Sichuan Province, also represents an early stage of the Ba Culture. Compared to the Shang dynasty period site of about four thousand years ago at Panlongcheng in present-day Hunan Province, Lijiaba is older, probably dating from the Xia dynasty (2205?–1766? B.C.). Many of the sites of the Ba people are located within Xiling Gorge, but later, about the time of the Shang, the center of the Ba Culture shifted to the area of Wu Gorge. Right now, outside of Shuangyantang, the three largest known sites in the Three Gorges area are: Lijiaba, Ganjinggou in Zhongxian, and Xiaotianxi in Fuling [all three in Sichuan Province]. The last site probably dates from the Warring States to early Qin dynasty period. Xiaotianxi is where tombs belonging to Ba aristocrats have been found.

According to ancient texts, the origin of the Ba Culture lies in Zhonglishan, in Wuluo along a southern tributary of the Yangtze known as the Qing in Hubei Province. From our recent investigations, we have learned about several important sites of the Ba Culture and have discov-ered its center, which lies in the region from Wushan to Yunyang. The culturally advanced Chu people living in Hubei blocked the Ba people from expanding eastward. Instead, the Ba were forced to retreat westward and made a living by fishing and hunting in the Three Gorges area and further west.

DQ: According to the 1982 Cultural Antiquities Law, antiquities graded 1, 2, or 3 (as established by the State Bureau of Cultural Antiqui-ties) are considered "precious." Although the 1987 Ministry of Culture circular "The Ranking of Cultural Antiquities" purports to provide criteria by which cultural antiquities are ranked, in reality, the determination of grades is a subjective decision made by a panel of the State Bureau of Cultural Antiquities or local cultural antiquities bureaus. The ranking is then approved by the State Council.* Are these sites which you have been discussing ranked as Grade 1, 2, or 3 and, therefore, protected at the national level?**

YW: Baiheliang [White Crane Ridge] has been declared a national-level site. But the other sites that we have discussed were only recently unearthed and are awaiting recognition. Most archaeologists believe that the earliest phase of a culture should have national protection.

Over the past forty years, the state has at different intervals recognized three grades of national-level relics and antiquities. A fourth grade is now being considered. The four or five sites within the Three Gorges area have yet to receive final approval by the State Council. The difficulty at present is that once the approval is issued, considerable efforts must be made to preserve and protect the sites. But if they are beyond protection, then these approvals are worthless. If the day after a site is approved as a national-level site it is submerged by a reservoir, then why go to all the trouble? Of course, we archaeologists will continue with our work as usual.

DQ: How is it that such invaluable sites have but one future, namely, inundation? Is the phrase "national-level antiquities preservation" (*guojiaji wenwu baohu*) nothing but empty words? How, as it is often said, do you "continue with your work as usual"?

YW: Let me tell you how I continue. Since last November we have been working on our proposal. After many discussions, the Three Gorges Construction Committee finally agreed to provide ¥10 million for preservation work. So far, however, the Yangtze River Planning Commission (*Changweihui*) has only allocated ¥2 million of that amount.

Over two hundred specialists are currently working in the Three Gorges area. These include over forty scientists from the Ancient Vertebrae and Ancient Humankind and Animal Research Institute of the Chinese Academy of Sciences, and this is the institute's largest site project since its establishment in 1949. Unfortunately, I had only ¥300,000 for

---

*See, J. David Murphy, *Plunder and Preservation: Cultural Property, Law and Practice in the People's Republic of China* (Oxford: Oxford University Press, 1995), p. 60.

**Since 1949, archaeologists in China have had a very difficult time gaining government protection for the country's vast storehouse of cultural antiquities largely because of political turmoil (during the Cultural Revolution rampaging Red Guards were encouraged by Mao Zedong to destroy ancient artifacts as they symbolized China's "old society"), general neglect, and government disinterest.

the whole lot, including the costs of boat rentals and local laborers. This was totally inadequate. University professors visiting the site could only afford to stay in shabby hostels, and the working conditions were extremely difficult. Nevertheless, after only a few months, the institute's basic research was completed. Some of the participants paid their own way and now they are asking me for reimbursement. I had no choice but to borrow ¥2 million from the Palace Museum in Beijing on my personal authority even though the money should have been allocated by the Yangtze River Planning Commission. I am at a loss as to why they refuse to give us the money.

DQ: As far as I know, no separate budget exists for the preservation of historical relics and cultural antiquities in the Three Gorges area. However, according to the official proposal, if money is badly needed it can be squeezed out of the population resettlement budget. Is this the conventional way of doing things in large-scale hydroelectric projects?

YW: According to international standards, the budget for the preservation of historical relics and cultural antiquities should be about 3 to 5 percent of the total project budget. In the mid-1980s, I worked with some Canadian experts from whom I learned of this standard. Two years ago when the National People's Congress formally approved the construction of the Three Gorges project, the total cost was put at ¥57 billion. The State Cultural Antiquities Bureau did a rough calculation based on the 3 to 5 percent standard and estimated that ¥1.7 billion was needed for the preservation work, and the Yangtze River Planning Commission acknowledged the accuracy of this estimate. Recently, however, the budget for the entire dam project has grown to ¥120 billion. Again, if we follow the 3 to 5 percent standard, the budget for relics and antiquities preservation should now be around ¥4 to ¥5 billion.*

DQ: Has the money been authorized?

YW: We have been given a verbal commitment that the budget for population resettlement in Hubei and Sichuan provinces will be ¥40 billion, and that this includes funds for the study, preservation, and protection of cultural antiquities. We have also been told that the total budget for antiquities preservation will at most be around ¥400 to ¥500 million and to forget about the 3 to 5 percent standard. For the next ten years or so, this is all the money we will have, and it will only be allocated to us after

---

*The most pessimistic scenarios for the ultimate cost of the Three Gorges project run to over ¥600 billion (U.S.$75 billion).

contracts are signed. But right now the costs for planning and the proposals for carrying out preservation work come from borrowed money.

DQ: Four or five hundred million yuan comes to only about 1 to 1.25 percent of the ¥40 billion for resettlement. Realistically, what can be accomplished with such a paltry sum? The removal of the Abu-Simbel temple at the Aswan dam in Egypt cost U.S.$40 million alone.

YW: I'm trying my best to work this thing out. Right now, we have identified over 1,000 sites in need of protection, but depending on how much money we get, we will have to choose only a few.* Of course, there is another problem too—even if we have enough money, we don't have enough experts to complete such a big project in such a short period of time.

For instance, how long will it take to properly excavate a site as massive as Shuangyantang? According to conventional practice, and with the manpower and material resources available to us, this one site would take several decades to excavate. But we have to have everything completed in ten years time, because at that point the area will be submerged. This requires more manpower, more money, and more advanced technology than we have. I have visited several subsurface sites being excavated by German scientists where the details of the sites—the layout of roads and streets, for instance—were explored without excavation and using new technology. Once the specific details of a site are known, then decisions can be made about which particular sites should be excavated. This is the only way.

DQ: Even if you were able to use this new equipment and technology on sites in the Three Gorges area, would you have enough time to excavate and relocate them?

YW: Absolutely not. No country in the world would dare to undertake such a large-scale excavation project in only ten years. We have no choice but to sacrifice some sites. Nonetheless, we will certainly try our best to minimize the loss. Keep in mind that we have only been talking about one specific site, Shuangyantang. There are many others like it in the Three Gorges area, including over one hundred Ba Culture sites alone. And the westernmost site of the Chu Culture is in Yunyang and it must be excavated. . . . How could such important sites be inundated?

---

*A list of priority-level cultural antiquities found in the Three Gorges area is found in Appendix D. A list of 16 archaeological sites which are likely to be inundated in 1997 is found in Appendix E.

A Yuan Dynasty (1271–1368) Buddhist shrine to be inundated.
(Photo courtesy of Elizabeth Childs-Johnson)

DQ: These are subsurface sites. What about those above ground? Are there any major problems in preserving these sites?

YW: In terms of ancient architecture, there are three major sites: Baiheliang, Shibaozhai, and Zhang Fei Temple. With a total budget of only ¥500 million, we will not be able to preserve Baiheliang. Because relocation is impossible, the only way to preserve it will be to construct an underwater museum. Li Tieying [a State Council minister] first made this proposal at a dinner with archaeologists. But unless the money is made available right away, Baiheliang will never be preserved. As time passes, the cost of building the underwater museum will rise substantially. The design of the underwater museum has been entrusted to Tianjin University.

The earliest known low water calligraphy is located at Lingshi, Chaotianmen. The difficulty with this particular site is that it is located on a navigation route and, therefore, in order to excavate and relocate it, navigation along the river would have to be halted. The cost, according to initial calculations, would be around ¥100 million.

DQ: What about folk culture, which is generally considered by archaeologists as "the living fossil of ancient civilization"?

YW: With or without the Three Gorges project, folk cultures in the area are on the verge of extinction. For example, at Dachang all the ancient houses and buildings are privately owned and are not considered historical relics. The owners can sell or tear down the buildings at will. Within the last year alone, the situation at Dachang has deteriorated significantly. If government budgets permitted, the simplest and most effective way to preserve these buildings would be to buy them from the private owners.

DQ: Ancient civilization belongs to all of humankind. Could overseas assistance help solve the budget problem—perhaps from Taiwan, Hong Kong, or other overseas Chinese communities?

YW: Indeed. We archaeologists endorse the concept of international cooperation, not just in terms of money, but also in terms of manpower. However, the office of the Three Gorges Construction Committee has repeatedly admonished us, telling us that we do not need outside assistance. A few days ago, an American newspaper published an article on the problems confronting the Yichang City Museum.* The interview was with Yao Yingqin, director of the Archaeology Museum, which is located near the dam site. The authorities came down and interrogated us: "Why did you mention those things to foreigners?" they asked. We told them that it was people from the Yichang Museum who had done all the talking. But a high-level official said to me: "We are capable of building the Three Gorges dam, so how can it be said that we can't come up with the money to preserve cultural antiquities and works of art?" He added that this was not his personal opinion but reflected the basic policy line laid down by the Three Gorges Construction Committee, namely: "Never discuss the issue of international assistance without our approval."

DQ: If the design of the dam were to be altered, for instance if its height were reduced from 175 meters to 150 meters, what would this mean for the preservation of cultural antiquities?

YW: Of course, the losses would be fewer.

DQ: And if the project's completion was postponed until the 2020s, would that provide more time for archaeologists to complete their work?

YW: Of course, this would make us more than happy.

DQ: As an expert who has been involved in historical and archaeological research for over forty years, what are your expectations and thoughts about the current situation?

YW: Archaeological work in China has certainly improved a great deal

---

*Philip Shenon, "Digging Up the Ancient Past, Before the Deluge," *New York Times,* September 10, 1994. Since the interview the museum has been closed.

over the last forty years. We are capable of doing a first-rate job in preserving the historical relics and cultural antiquities of the Three Gorges area. As for expectations, as an archaeologist and the person responsible for cultural antiquities preservation in the Three Gorges area, I sincerely hope that the Three Gorges Construction Committee carries out its work in a steadfast fashion and that it follows international standards. Moreover, in order to protect the cultural antiquities, we should seek international support both in terms of money and manpower. Personally, I am someone with the highest national and ethnic pride, and so I do not wish to see China lose face on this issue. On this particular matter, however, we must seek international assistance. What needs to be done should be done.

Second Interview, August 1995

DQ: You mentioned in our first interview that beginning in November 1993 scholars involved in the preliminary work to assess the protection of historical relics and cultural antiquities in the Three Gorges area were forced to "pay their own way." It has now been eight months since the formal inauguration of the Three Gorges project in December 1994. Have you received the funds promised to you by the state to implement your preliminary assessment?

YW: Implementation of the assessment plan began in November 1993, and has involved over 20 major academic units from all over the country. These included the Chinese Academy of Sciences, the Chinese Academy of Social Sciences, and Beijing, Qinghua, and Tianjin universities [China's three premier institutions of higher learning]. In March 1994, I was formally appointed the director of this effort. My first act was to negotiate funding with the Three Gorges Construction Committee but nothing really came of it until April 1994, several months before the formal inauguration of the dam project in December. At that point, I held several meetings with representatives from the Resettlement Bureau of the Three Gorges Construction Committee and the population resettlement bureaus of the Hubei and Sichuan provincial governments. On April 31, 1995, several months after the project began, I signed an agreement with these organizations, and in early May we finally received approval of our funding as stipulated in the agreement to initiate the assessment plan.

DQ: Thank God. It seems that Li Boning, who was in charge of resettlement and the protection of cultural antiquities for the Three Gorges

Construction Committee, and his successor, Tang Zhanghui, have finally come to their senses, and realized that this is no trivial matter. Is that true?

YW: Do you see it that way? Let me tell you. We started talking about funding last April, but I didn't receive anything until the National People's Congress (NPC) and the Chinese People's Political Consultative Conference (CPPCC) met in April of this year. Prior to those gatherings, I had convened my own meeting where members of my group reported to a select number of NPC and CPPCC delegates on the development and implementation of the assessment plan. After delivering our report, everyone there shouted at us: "Such an important matter and you have been conducting this behind closed doors. Why didn't you scream and yell at the Three Gorges Construction Committee and the State Council?" Others said: "In 1992, when the NPC was considering the total budget of the Three Gorges dam, why was funding for the protection of cultural antiquities not included?"

During these two meetings a series of proposals were worked out, which journalists then reported on in internal reference* and public newspapers. The proposals advocated that funds be allocated immediately to implement the assessment, criticized the funds provided in the agreement as woefully inadequate, and recognized that this task should be managed by the central, and not the provincial, government. The curator of the Beijing Municipal Library, Ren Jiyu, joined in by writing a letter and meeting personally with Zou Jiahua [a vice premier], at which time Ren proposed that the funds earmarked for the protection of cultural antiquities not be part of the resettlement budget but be separate. I also wrote a letter to Zou in which I stressed that if the funds to carry out the assessment were not authorized soon, we would have to terminate our work. Zou then gave his approval to Guo Shuyan who, in turn, gave the green light to his subordinates resulting in an executive order to release the funds in late April finally being issued. Getting these funds was, therefore, the result of a collective effort.

DQ: Is it true that your work is almost completed—that you have determined which sites should be protected and relocated and which will be submerged by the reservoir?

YW: Between August and October 1995, a plan for 22 counties will be developed and submitted for approval. By the end of December, planning

---

*Publications restricted to top-level leaders in China.

for both Hubei and Sichuan provinces should be completed. By March 1996, an overall plan should be completed, consisting of 25 separate reports. As for the total number of sites, our investigation suggests that there are 1,271. We must be very careful in our estimates about the total number of sites which must be protected. I would say a conservative figure is about four hundred to five hundred.

DQ: What about manpower? And the funding problem? As I recall, in 1991 the Yangtze River Planning Commission (under the authority of the Ministry of Water Resources) stated that the Three Gorges project would submerge between 60 and 70 cultural sites, and that the estimated cost of protection was ¥60 million. Are these figures still valid?

YW: Those are the figures proposed by the committee in 1990. I have no idea how it arrived at them. In November 1994, the State Council decided that funds for the protection of cultural antiquities would be taken out of the ¥40 billion population resettlement budget. The budget for cultural antiquities would, therefore, come to around ¥300 million. This figure is still valid.

DQ: Is that possible? As you mentioned in our last interview, the general rule for calculating a budget for preservation of cultural antiquities should be about 3 to 5 percent of the budget for the entire project. Based on a conservative estimate of the project's total cost (¥120 billion), the budget for the preservation of cultural antiquities should be around ¥4 to ¥5 billion. Why only ¥300 million? I should point out that the cost of relocating structures from the Central Fine Arts Academy [in Beijing] alone cost one billion yuan! Three hundred million yuan for the entire Three Gorges area! Are you kidding! Did you agree to this by signing the resettlement budget you just mentioned?

YW: That figure is from Item Number 10 in the document titled "Professional Project Restoration and Rebuilding Compensation Investment" from a report on the estimated cost of compensation for resettlement in the Three Gorges area that was issued at the end of last year. Let me tell you a story. A few days before signing the budget agreement, I had a conversation with a high-level official responsible for resettlement in the Three Gorges project. I said to him: "Now that we're alone, tell me frankly, is ¥40 billion enough for population resettlement?" He smiled but didn't say a word. Then I told him: "Although the State Council has decided on a budget, I know for sure that ¥40 billion is not enough. Let me tell you something else. Three hundred million yuan for cultural antiquities from that ¥40 billion is inadequate! The only difference between the two of us is that you're not intimidated by the inadequacy of the ¥40 billion resettlement budget because you know there are a million peasants out there whom you know will protest when the time comes. We archaeologists can only clam up no matter how dire the situation."

He then replied: "You must understand. The State Council requires us to use limited funds and work out a plan. I am in no position to comment on any of this. The budgetary agreement signed by our Resettlement Bureau and your cultural antiquities protection group is based on that allocation. I cannot say whether it can be changed."

I responded: "We are from the academic realm. It is up to the government to decide on budgetary allocations. Our job is to provide an accurate figure for the actual amount of money needed. As an academic, I must be responsible for my work. I cannot concoct a figure based on how much money is available from the government simply so the Three Gorges project can be launched. All I can say is that the ¥300 million for antiquities protection is far from enough and, therefore, I can't sign the agreement."

Again he responded: "What if the authorities require you to sign? What will you do?" I said: "Fine, I'll resign and you can look for somebody else." In the end, he said: "Let's make a compromise. Try to conduct your work with the amount given. You can submit proposals later for work that requires more money." And so I signed the agreement, but there is a clause in it that says: "The budget for cultural antiquities protection should be based on actual need."

DQ: Did the agreement specify any limits on additional monies for the protection of cultural antiquities?

YW: No. The understanding was that the agreement called for providing whatever will be needed. Now that we have received the money to fully execute our initial assessment, by early next year we'll be able to provide an outline for the overall plan to relocate the cultural antiquities. The State Council will invite other departments to assess and improve on our plan. Personally, I think whatever the government provides will be far less than what is needed. We got into a big argument with the Yangtze River Planning Commission over the whole issue of the protection of cultural antiquities. I said that if they asked for our advice, it would be that if the Three Gorges weren't built, then protection of cultural antiquities wouldn't be necessary. This would be the best of all possible worlds. However, now that the project has been launched, we will try our best to save as many historical relics and cultural antiquities as possible. It's not a matter of asking for a certain sum of money. Even if adequate funding is forthcoming, it will only mean that we can rescue an additional 10 to 20 percent of the relics. The majority will still be submerged. At this point, with the project moving along, and the cultural antiquities in need of protection, the only solution is to call for international support.*

DQ: The problem is that the project has already started. Can the relics be excavated while the project is being built?

YW: Protecting cultural antiquities is not like digging for sweet potatoes, you know.** It's not as if once you dig them out the work is finished. Such excavations require meticulous scientific recording and treatment. The problem is that even if we do receive the required amount in line with international standards (that is, 3 to 5 percent of the total budget), it is still inadequate. We are trying to minimize the losses as much as possible, but at most we will succeed in preserving about 10 to 20 percent of the sites we found in our assessment.

DQ: Could you seek support from UNESCO? Wasn't the preservation of cultural antiquities at the Aswan dam project in Egypt done with international cooperation, implemented through UNESCO?

YW: Every time someone mentions Aswan, the Three Gorges Construction Committee says: "Ours is a different situation." I don't know why, but they are deeply reluctant to talk about international support. They said to us: "We haven't asked for overseas support and you shouldn't either. It's up to us to decide. The Three Gorges project is unique, and even the State Cultural Antiquities Bureau has no right to seek international assistance." As far as we're concerned, however, the two most difficult problems facing us right now—money and manpower—require us to call for international support. Even if the government eventually provides the necessary funds, it will be impossible to complete the massive amount of work required in such a short period of time, even if we mobilize every archaeologist in the country.

There is the possibility of help from overseas Chinese in Taiwan, Hong Kong, and Macao. Last November, Zhang Dele from the State Cultural Antiquities Bureau and I headed a Chinese museum delegation to Taiwan. When the protection of cultural antiquities around the Three Gorges was

---

*In August 1996, Yu Weichao and 55 other prominent academics and officials took the unusual step of writing to Jiang Zemin to advocate that preservation work be sped up and adequately funded. That letter is found in Appendix F.

**A sarcastic swipe by Yu Weichao at the general ignorance of China's official cadres and decision makers on matters such as archaeology.

raised, I said that we would willingly accept help from organizations in Taiwan such as Academia Sinica, Taiwan University, and the Natural History Museum in Taizhong City. Of course, the costs would have to be borne by the Taiwanese. I also told them that whenever two or more identical pieces were excavated they could be shared between Taiwan and the mainland. In the case of Hong Kong, I met with the historian Xu Zhuoyun, who was very excited about the prospect of helping to raise money from foundations in Hong Kong.

DQ: What about Singapore?

YW: We haven't made any suggestions on that matter. I would like to establish a Three Gorges Cultural Antiquities Protection Foundation that would operate as a nongovernmental organization, and I hope that the Chinese government will go along with this idea. Protecting cultural antiquities enhances everyone's understanding of ancient cultures. This is not a matter that China should consider relevant to its national security. We should discourage narrow, parochial nationalism and invite scholars from all over the world to join in the effort.

In addition to our preservation work, we should take advantage of this opportunity to conduct research on these ancient cultures. I personally would like to get in touch with international networks and organizations. The people of China have not adequately addressed how we lag behind the rest of the world in the humanities. Nor has our national leader [Deng Xiaoping] acknowledged it. He does not understand that it is impossible for a country to gain national independence through technology alone. Without a concern for the humanities, it will be impossible to regulate society properly, which will ultimately hinder our economic development. Since the late 1960s, research on ancient cultures has made great strides, especially in archaeology and anthropology. We must, therefore, get in touch with the world. Otherwise, we will fall further and further behind.

DQ: Perhaps this idea of yours will be very hard for others to accept. The loss of status and a role for the humanities and social sciences in China is nothing new. It occurred many years ago.

YW: Personally, I think that it is possible for us to use the preservation work in areas such as the Three Gorges to increase our understanding of humanity and culture. For instance, the Ba Culture has been directly linked by a consanguineous relationship (*xueyuan*) to the contemporary Tujia minority group. These two peoples live in roughly the same area around the Three Gorges in Hubei and Sichuan provinces. I sincerely hope that our archaeological research can be linked to the study of ethnicity and folk customs so that valuable comparisons can be made. One is a cultural comparison between relics from three to four thousand years ago and those still in use today. A second comparison could involve DNA. From the sites along the Three Gorges area we could excavate tombs of the Ba Culture that date as far back as the Han dynasty and even the Western Zhou (1100–771 B.C.). We could then compare the DNA from these two-thousand-year-old bones with that of the Tujia, in effect carrying out a comparative study between two cultures. The sites in this area are too unique and valuable to be lost.

In short, an enormous challenge awaits the scholars involved in excavating and preserving the cultural antiquities which are to be submerged.

Chapter Ten

A Lamentation for the Yellow River

The Three Gate Gorge Dam (Sanmenxia)

Shang Wei

"A Clear Yellow River"

In August 1949, just before the establishment of the People's Republic, a "Preliminary Report on Harnessing the Yellow River" (*Zhili Huanghe chubu yijian*) was delivered to Dong Biwu, chairman of the North China People's Government of the Chinese Communist Party (CCP). One of the report's authors, Wang Huayun, headed the Yellow River Research Group and would later serve as director of the Yellow River Commission (*Huang weihui*) which was charged with advising the Ministry of Water Resources and Electric Power on management of the river's resources.* Comrade Wang was known as the "leading expert" on harnessing the Yellow River, and he has also come to be known as one of the founders of the Three Gate Gorge dam (*Sanmenxia*) project located in Henan Province.

According to the 1949 preliminary report, controlling the perennial floods on the lower reaches of the Yellow River required that a number of dams and reservoirs be built. However, the question of where these dams

---

*Such commissions exist for all the major rivers in China and are charged with advising the Ministry of Water Resources and Electric Power on management of river resources. But since they have no authority to issue orders that fall within their jurisdictions, it is virtually impossible to implement integrated plans for river-basin development. See, Lieberthal, *Governing China* (New York: Norton, 1995); p. 286.

and reservoirs would be built remained unanswered. In the report, Wang asserted that there were three possible sites between Shan County and Mengjin City—Three Gate Gorge, Balihutong, and Xiaolangdi. This proposal marked the beginning of serious planning for a dam at the Three Gate Gorge.

In spring 1950, the Yellow River Commission [now led by Wang] completed another study of prospective dams on the Yellow River. The commission report concluded that the benefits that would accrue from developing a dam at Balihutong or Xiaolangdi were not great, while a dam at the Three Gate Gorge would flood one million people from their homes, an issue that "deserves considerable attention," according to the report.[1]

In summer 1950, Fu Zuoyi, then minister of water resources, headed a delegation to the Soviet Union to study water resource development. The delegation included Zhang Hanying, then vice minister of the Ministry of Water Resources, and Zhang Guang, who would later play a critical role in promoting the construction of the Three Gate Gorge dam. Upon the delegation's return, Fu Zuoyi delivered a report on developing water resources to the State Council. In tone and in substance, Fu argued for a thorough and comprehensive approach to developing the Yellow River. He urged that the banks and dikes of the Yellow River be reinforced, that agriculture and forestry policies be integrated to improve soil conservation on both the mainstream and the tributaries, and that preparations be made for the construction of a reservoir between Tongguan pass in Shaanxi and Mengjin to block flood waters and sedimentation in the tributaries. The emphasis was clearly on damming the Yellow River's tributaries, since, according to Fu, building a dam on the mainstream posed serious political, economic, and technical problems.[2]

Although Wang Huayun's Yellow River Commission had, in its 1950 report, made note of the scale of resettlement required if the Three Gate Gorge site were chosen, it did not favor damming the tributaries as Fu suggested. The commission claimed that there were too many of them, that the flood control benefits were unreliable, and that the cost was prohibitive. Thus, yet another commission study was prepared for the project, this one more positive than its 1950 predecessor. The new study made sensational claims about the prospective flood control, electricity generation, and irrigation benefits of a dam at the Three Gate Gorge site and proposed the construction of a large dam with a normal pool level of 350 meters that would both store flood waters and block sedimentation behind the dam. At first glance, it seemed that this plan would put an end to flooding in the lower reaches caused by

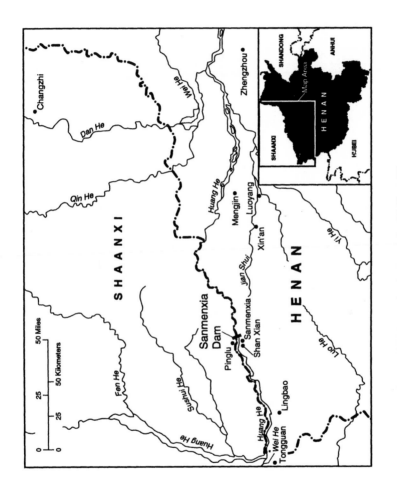

The Sanmenxia dam on the Yellow River.

the river's high sediment load.* But where to put the sediment? The answer, of course, was in a reservoir large enough to store it.

In spring 1952, Wang Huayun accompanied a group of Soviet engineers to the Three Gate Gorge site. The Soviets, who had been building grandiose dam projects in Siberia for decades, made fast friends with the members of the Yellow River Commission.[3] Out of this visit a high dam was once again proposed for the Three Gate Gorge area. But the problems related to resettlement and the inundation of farmland remained, and dam opponents drew attention to them, stalling the project.

Less than a year later, the project began to gain momentum again. In October 1952, Mao took a tour of the Yellow River and uttered a line which resonated with both sides in the controversy over whether to build the Three Gate Gorge dam. He said: "Work on the Yellow River must be carried out well." But *what* was to be carried out well? A large dam? Smaller dams on the tributaries? Both sides tried to claim Mao's words as support for their views, but, in the end, the pro-dam side won the struggle. Shortly thereafter, the Yellow River Commission rejected, once and for all, the idea of building a series of smaller dams on the tributaries and insisted on the large-scale Three Gate Gorge dam. Wang Huayun personally submitted a report to Mao to this effect.

The dam opponents were reduced to trying to block the project purely on financial terms. Deng Zihui, the head of the CCP's Rural Work Department, did his part by authorizing a "miracle" budget of a mere ¥500 million, which would pay for the resettlement of only 50,000 of the 600,000 to one million relocatees who would be displaced by the dam. Dam critics then peppered supporters with questions about how they planned to properly resettle people with so little money. By forcing pro-dam officials to admit that the budget was woefully inadequate, the anti-dam forces hoped to kill the project. But the plan failed. When asked about resettlement, dam supporters would quickly shift gears and ask, "Well what about flood control?" As Zhou Enlai commented: "They seemed to believe that if the Three Gate Gorge dam was not built, the lower reaches of the Yellow River would be immediately devastated by flood."[4]

Deng Zihui had no choice but to try to personally persuade Mao to oppose the project.** He told Mao that he basically agreed with the mea-

---

*In the early 1950s, the Yellow River carried about 1.3 billion tons of silt through the Three Gate Gorge annually. See Smil, *The Bad Earth*, p. 44.

**Mao Zedong's critical role in playing off factions and arbitrating all such disputes in the "court politics" of the 1950s is discussed in Frederick C. Teiwes, *Politics at Mao's*

sures for flood control. But rather than build the large dam at the Three Gate Gorge, he argued that during the First Five-Year Plan [1953–57],* the government should build two reservoirs at Zhichuan and Mang hill near Luoyang. Then, after five or ten years, the country would have the resources and technical expertise to build large-scale reservoirs and deal with large numbers of relocatees. Although this was not exactly what Mao wanted to hear, he could not refute what Deng was saying. And so, once again, the project was postponed.

In January 1954, a contingent of Soviets known as the Yellow River Planning Soviet Experts Group came to Beijing at the invitation of the Chinese government. From February to June 1954, a 120–member investigation group composed of both Chinese and Soviet experts surveyed the area from Liujia Gorge to the mouth of the Yellow River. They concluded that Three Gate Gorge was a good site for a dam project. This opinion played a decisive role in the decision to launch the project, and Deng Zihui was powerless to oppose it.

During the survey, in April 1954, the central government decided to entrust the design of the dam and reservoir to the Soviet experts. Their design, delivered eight months later, recommended a 350–meter normal pool level and a reservoir with a 36 billion cubic meter storage capacity. The primary purposes of the dam would be to:

- reduce the 37,000 cubic meters per second flow of a 1,000–year-frequency flood in the upper reaches of the river to 8,000 cubic meters per second so that the danger of floods would be dramatically reduced;
- accumulate all sedimentation from the upper reaches and release only clear water to the lower reaches so as to realize the long-held goal of "a clear Yellow River," and to prevent the riverbed of the lower reaches from continually rising because of accumulated sediment;**
- adjust the quantity of water in the Yellow River and to irrigate

---

Court: Gao Gang and Party Factionalism in the Early 1950s (Armonk, N.Y.: M.E. Sharpe, Inc., 1990).

*The major slogan of the First Five-Year Plan was "Let's be modern and Soviet." See Riskin, *China's Political Economy*, p. 53.

**In Henan and Shandong provinces where it is confined by dikes, the Yellow River flows three to five meters, and in some places up to ten meters above the surrounding countryside. See Smil, *The Bad Earth*, p. 47.

initially 22,200,000 mu of farmland (to eventually be raised to 75,000,000 mu).

- install an electrical generating capacity of 900,000 kilowatts and annual electricity generation of 460 million kilowatt-hours; and
- improve navigation on the lower reaches.

In its conclusion, the Soviet design report suggested that the project would bring about enormous benefits. However, two serious problems remained: First, the project would submerge 2.07 million mu of land and force 600,000 people from their homes. And, second, because of the accumulation of sediment in the reservoir, the dam would only last twenty-five to thirty years. Thus, debate over the project shifted from whether it should be built at all, to how to reduce the accumulation of sediment in the upper reaches.

The pro-dam forces called for extensive soil conservation in the watershed, a proposal that would ultimately play a decisive role in the final decision to launch the project. In a "technical task report" produced by the Chinese contingent of the joint Soviet and Chinese experts' group, new data on sedimentation was released: By 1967, the Chinese claimed, the sediment delivery to the reservoir would be reduced by 50 percent, and after fifty years it would be reduced by 100 percent.*[5]

The Soviet contingent of the group could not tolerate such bombast and urged caution in estimating the benefits of soil conservation measures. They concluded that "by 1967 the sediment would be reduced by 20 percent and in fifty years time by 50 percent."[6] Despite the Soviet experts' concerns, however, the project was now well on its way.

At the second plenum of the First National People's Congress in summer 1955, Deng Zihui delivered a "Report on the Summary Plan for Controlling Yellow River Floods and Opening Yellow River Irrigation Projects" on behalf of the State Council.** The report was passed unani-

---

*While soil conservation measures such as reforestation can significantly reduce the suspended sediment conveyed by the river into the reservoir, it is a practical impossibility to reduce the amount by 100 percent. All rivers, even in the best protected watersheds, convey appreciable amounts of sediment during natural floods.

**This grandiose plan for the utilization of the Yellow River had as its centerpiece the construction of 46 separate dams on the river to impound water for the generation of electricity, expansion of the irrigated areas in the river basin, and increased navigation. The dam at the Three Gate Gorge was to be the largest of the 46. See Smil, *The Bad Earth*, p. 45.

mously, but Deng was soon ousted from his position. This prompted Zhou Enlai to say to Deng: "You gave such a bold and frank speech on the various problems posed by this plan that you're now a household name."[7]

Later, the CCP decided to entrust further design work for the dam to the Leningrad branch of the Soviet Power Stations Water and Electricity Design Institute, but stipulated certain points in the design. The Soviet designers were told to:

- extend the lifespan of the reservoir;
- set the normal pool level between 350 and 370 meters and design a separate plan for every five meters in between;
- ensure flood control safety by making sure that the allowed water release amount can be reduced from 8,000 cubic meters per second to 6,000 cubic meters per second; and
- consider the possibility of extending the amount of farmland to be irrigated.

In April 1956, the Soviets delivered their revised design which set the normal pool level at 360 meters and included an alternative 370–meter plan which would insure a 100–year lifespan for the reservoir. In July 1956, China's State Capital Construction Commission approved the 360–meter design. The design did not, however, solve any of the major problems that had been identified earlier; the dam and reservoir would submerge 3.33 million mu of land and force the relocation of 900,000 people. Naturally, Shaanxi Province, where much of the reservoir and many of the relocatees were located, found this decision difficult to swallow. But the pro-dam forces were overjoyed; they established the "Three Gate Gorge Dam Project Bureau" (*Sanmenxia gongchengju*) even before the design was given final approved by Beijing.[8]

Amid all of the enthusiasm and chaos there appeared two intellectuals who would play key roles in the evolution of the project. One was a professor from a renowned university, and the other was a young college graduate. They disagreed with the plans to build a high dam at the Three Gate Gorge site and calmly explained their views to the top decision makers. But they were ignored. When I met with them in 1991, their lowly political and professional status had not changed for over thirty years. The professor was named Huang Wanli, and the young engineer, Wen Shanzhang.

The Specialists

The Three Gate Gorge project was officially launched on April 13, 1957. Despite the years of study, construction began without final designs for key elements of the project, including the normal pool level, the level of dead storage (*si shuiwei*), and the size of the gates for periodic release of flood waters. Moreover, basic decisions about whether the dam would be used primarily for flood control or for electricity generation, the amount of flood water it could store, and whether sediment should be blocked behind the dam had still not been made.* But the project went ahead at full speed anyway.

Huang Wanli was the son of the CCP's ally Huang Yanpei. At the time of this controversy, Huang Wanli was a professor in the Department of Hydrology at Qinghua University, China's premier technical institution. Not yet fifty years old, he was a well-educated and experienced engineer. After graduating from Tangshan Communications University in 1932, Huang became a structural civil engineer. But after flooding of the Han River [a tributary of the Yellow River] in 1931 and the collapse of the Yellow River dam in 1933, many young people became interested in studying hydrology.[9] In fact, a survey of Chinese engineers conducted after the Han River flood revealed that all of China's hydro engineers were civil engineers. There was not a single hydrological engineer among them. This meant that the engineers could design dams, but knew nothing about river hydrology. Huang therefore decided to make hydrology his specialty.

In January 1934, he went to the United States. Assuming that floods occurred after rainstorms, he decided to major in meteorology. After completing a master's degree at Cornell University, Huang shifted his focus to Geology for his Ph.D. at the University of Illinois. After receiving his doctorate, Huang set out to visit every major hydro project in the United States, which he followed with a four-month internship at the Tennessee Valley Authority [now the United States' largest public utility]. On his return to China, Huang was appointed a hydro supervisor for the Economic Committee of the Kuomintang government. Later, he became an engineer and survey director for the Bureau of Hydrology in Sichuan Province. During the War of Resistance against Japan [1937–45], he sur-

---

*Dam designers optimistically asserted that " 'difficulties that may arise in power generation, irrigation, and navigation as a result of the silting up of the reservoir . . . will be comparatively easy to deal with.' " Quoted in Smil, *The Bad Earth*, p. 45.

veyed the Min, Wu, and Jialing rivers on foot. In 1947, he became director of the Gansu Province Hydrological Bureau, and, in 1949, was appointed as a consultant to the Northeast Hydro Bureau. According to Huang: "The study of geography and geology was an eye-opening experience that made me realize that the structural theories of civil engineering were inadequate for solving the problems of floods. Dam projects are built on rivers and they change the dynamics of water flow which, in turn, cause changes in the riverbed. When I first returned to China, the field of geology had yet to be established. Ten years later, I had walked 3,000 kilometers and collected rich data on various aspects of hydrology and information on the harnessing of rivers."[10]

After hearing about the Soviets' December 1954 Design Report, and the 1956 Revised Design which followed, Huang Wanli made a submission to the Yellow River Commission, in which he raised doubts about the entire planning procedure.

Although Huang believed that dams could effectively adjust the flow of water in a river, he felt they were bound to destroy the natural flow of sediment. He also felt that dire consequences would follow from the policy of keeping the lower reaches of the Yellow River "clear" and free of sediment. Huang argued that top soil on the banks of the river should remain where it was, while dirt or sediment in the riverbed must flow continuously. The idea of purposely accumulating sediment behind a dam was wrong, he argued. It was an attempt to fiddle around with the laws of nature.

Huang also argued that the accumulation of sediment would limit the reservoir's lifespan and, furthermore, that no one had bothered to consider what would happen to the upper and lower reaches when the dam was no longer in use. As a result of these and other concerns, Huang categorically opposed building large-scale dams on the mainstreams of major rivers. If a dam had to be built on a mainstream, Huang insisted that sediment be flushed out of the reservoir as much as possible [the exact opposite of the plan to eventually create a "clear Yellow River" downstream of the Three Gate Gorge dam]. To achieve this, Huang felt that large sediment sluice gates should be installed at the base of the dam.

Huang Wanli is a courageous man. For over forty years he was virtually alone in his opposition to the Three Gate Gorge project, and to the planned Three Gorges project on the mainstream of the Yangtze River. He has been publicly attacked and isolated for his views, but he has never compromised.

Wen Shanzhang is a hydrologist who studied in the Soviet Union with the Soviet delegation but disagreed with their proposal for the Three Gate Gorge dam. Instead, he suggested a lower dam (with a 335–meter normal pool level) and a small reservoir that would release flood waters and flush sediment. This dam would generate less electricity, but it would also cause less inundation, require the resettlement of fewer people, and be much cheaper to build.[11]

In June 1957, two months before the launch of the project, over 70 experts took part in the "Symposium on the Three Gate Gorge Pivotal Hydro Project."[12] Wen Shanzhang was the first to speak. Rather than suggest a few simple modifications to the official plan, Wen advocated the construction of a smaller dam, as described above. Above all, Wen argued against adopting a "grandiose" approach to the dam's construction.* Wen opposed the Soviet design and argued that it was not necessary to build a large-scale project like the Three Gate Gorge dam for irrigation alone. Moreover, he felt that eastern and northern Henan should develop their own irrigation networks and that irrigation needs in other areas could be met by building a series of smaller dams. He also explained that a smaller dam could successfully control floods and lower the cost of the dam by ¥450 million. It would also lower the number of relocatees to fewer than 150,000 and reduce the amount of land to be submerged to only 500,000 mu.

But Wen Shanzhang faced a wall of opposition. The highly respected Li Eding, who was later appointed chief engineer of the Three Gate Gorge project, and the Soviet-trained Shen Chonggang adamantly supported the original official plan, as did Li Bindu.

Li Fudu was the nephew of the father of hydrology [Li Yizhi] in China, and deputy director of the Yellow River Hydro Development Commission (*Huanghe shuili weiyuanhui*) in Henan Province. Just a week before the symposium, Li Fudu had also advocated the idea of building small-scale dams and relying on them and well water for irrigation, but at the symposium he changed his mind and supported the official 360–meter design. Li's speech was undoubtedly key to the final decision.

Huang Wanli responded by reiterating his claims that the extensive plans for soil conservation in the upper reaches would not effectively

---

*"Grandiose" plans still have considerable currency in China as is indicated by the Three Gorges dam and plans to irrigate China's deserts in the northwest by using nuclear explosions to cut canals through rugged mountains in order to bring water from Tibet.

control sedimentation, and that the goal of a "clear" Yellow River in the lower reaches distorted the laws of nature. How is it, Huang asked, "that you object to the accumulation of sedimentation below the dam but allow accumulation above it?"[13]

Zhang Shouyin (an engineer in the Hydrology Bureau of Shaanxi Province) chimed in on the issue of resettlement, explaining Shaanxi Province's opposition to the large-dam design. "When the water level is raised from 350 to 360 meters," he noted, "the number of relocatees increases sharply from 437,000 to 735,000."[14] Zhang therefore suggested that the normal pool level not exceed 350 meters; otherwise, he warned, sediment accumulation would lead to floods that would threaten the city of Xi'an.

The symposium ended on June 24, and a "comprehensive proposal" was approved. The official publication of the Ministry of Water Resources and Electric Power claimed that the proposal was approved by a great majority of the participants. But a memorandum issued by the symposium itself paints a different picture: Twelve people, or one-fifth of the symposium's participants, openly opposed the proposal.[15] The symposium's final design was approved by the State Council.[16]

Forced Reconstruction

One year after the project was launched, six million square meters of earth had been moved, and 30,000 cubic meters of reinforced concrete had been poured.[17] But Shaanxi Province still opposed the dam proposal and continued to request changes to it. In late 1958, after the disastrous Chengdu Conference [where rapid development of water resources was approved along with an initial proposal to construct a dam at the Three Gorges on the Yangtze], Zhou Enlai, Peng Dehuai, and Xi Zhongxun visited the Three Gate Gorge construction site. Both Peng and Xi had considerable political clout with the Northwest Bureau of the CCP, and it was understood that once they approved the project Shaanxi Province would have no say.

Although Zhou Enlai made no public statements about the project during this site visit, his doubts and skepticism were evident. One day, as charts and diagrams were being set up in a meeting room, Zhou walked in and asked why none of the charts included a dam design based on the 335–meter normal pool level [the design proposed by Wen Shanzhang]. The engineer in charge responded that the 335–meter proposal would create a reservoir insufficiently large, to which Zhou, in turn, responded:

"The only thing you people understand is large-scale this and big that."[18]

At a subsequent meeting held at the site, debate was heated. The Shaanxi delegation gave impassioned speeches requesting a lower normal pool level, and Zhou tried to support them. Xi Zhongxun proposed a compromise—a 360–meter design, an actual height of 350 meters, and effective water storage at 340 meters. The compromise was accepted, meaning that the pro-dam group's desire for a 360–meter dam was not completely rejected but that the normal pool level would, in reality, be only 340 meters.

By the end of 1958, the Yellow River had been blocked, and by 1960 the dam was 340 meters high, tall enough to hold back floods. In September of that year, water storage and sediment accumulation began.

But the dam supporters and Soviet experts had made a serious mistake. At the June 1957 symposium (discussed above), Huang Wanli had cautioned that the designers must include sediment sluice gates at the bottom of the dam and silt discharging tubes so that accumulated sediment could be flushed through the dam. The majority of those attending the symposium had agreed, and the initial design approved by the State Council had included the outlets and tubes. But under the leadership of Professor Zhang Guangdou, and based on suggestions from the Soviet experts, the 12 tubes specified by the initial design were blocked by reinforced concrete during construction. By 1967, each and every one had to be reopened at a cost of ¥10 million each.

In February 1962, the first 150,000 kilowatt turbine began operation. The project was completed on schedule, but there was little to celebrate. The reservoir caused farmland to become waterlogged and the agricultural output of nearby peasants dropped sharply because of increased salinization and alkalinization. Moreover, sediment began accumulating in the reservoir and upstream river channel immediately, and was soon inching its way upstream toward the industrial center of the city of Xi'an.* As a result, in March 1962, the project had to be extensively rebuilt and altered to release flood water and flush out sediment, thereby resulting in a drop in the normal pool level. With the drop in the reservoir level, the 150,000 kilowatt turbines quickly became useless and were moved to the Danjiangkou power station—a very costly operation and one that effectively removed all of the dam's electricity generating capacity for the time being.

---

*Between September 1960 and March 1962, 1.645 billion cubic meters of sediment entered the reservoir, 93 percent of which was "deposited" behind the dam. At this rate, the reservoir's life would be just a few years. See Smil, *The Bad Earth*, p. 46.

Despite the reconstruction, however, sediment continued to move toward the upper reaches. At the April 1962 third session of the Second National People's Congress, delegates from Shaanxi Province proposed lowering the normal pool level to below 315 meters and opening all of the locks to release flood waters and flush out sediment. In other words, they were proposing that the Yellow River flow freely, and that the dam become a run-of-the-river design.

By the third year of operation, the reservoir had accumulated so much sediment—approximately five billion tons—that the sediment tail was only 30 kilometers from Xi'an.* Shaanxi Province appealed directly to Mao Zedong. When faced with the reality of the situation, Mao became very upset and told Zhou: "If nothing works, then just blow up the dam."[19]

But destroying the dam was a difficult proposition at best, so Zhou Enlai convened a series of meetings to come up with an alternative plan. Between 1965 and 1968, the project was rebuilt again. The normal pool level was reduced to 315 meters, almost doubling the amount of water (and, hopefully, sediment) that the dam could release per second. But above Tongguan pass, the sediment continued to build up. The Yellow River simply would not behave itself.

Eventually, the dam, which was now full of holes, was able to release enough water and flush out enough sediment to balance further deposition, at least temporarily. In 1978, five turbines, with an installed capacity of 250,000 kilowatts, were installed at the dam. (The original design called for 1,160,000 kilowatts.)

Despite all of the problems with electricity generation and sedimentation, many felt that the effects of inundation and resettlement were even more serious. Based on the 360–meter design, 3,330,000 mu of land was to be submerged and 900,000 people were to be resettled. When Zhou Enlai had the normal pool level essentially reduced to 335 meters, the effects were reduced significantly—only 856,000 mu was submerged and 318,900 people resettled.** But many were resettled for naught—the project was such a failure that some relocatees' land was never submerged.[20]

---

*As water enters the reservoir it slows, depositing sediment in the river channel. Over time, the "sediment tail" or area of backwater effect moves progressively upstream from the reservoir.

**An additional 84,900 people had to be resettled as a result of a landslide in the area, bringing the total to 403,800.

Moreover, the mass of accumulated sediment created vast swaths of new land.

However, the new land was not given to the relocatees but was occupied by soldiers. When the relocatees learned of these facts they were shocked and immediately petitioned that they be allowed to return to the land and that the soldiers be vacated. In May 1985, (according to Central Document No. 29), 300,000 mu of land that had been occupied by soldiers and run as state farms was returned to its rightful owners, the relocatees.*

In the late 1980s, thirty years after the dam was built, sedimentation once again became a problem. This time the favored solution was the construction of another dam—the Xiaolangdi—in the lower reaches of the Yellow River. The new reservoir would have a total capacity of 12 billion cubic meters of water, would control floods, and would help reduce sedimentation in the lower reaches of the river for up to twenty years. After that new measures would have to be adopted.

During the thirtieth anniversary of the Three Gate Gorge dam the following refrain was repeated over and over:

> Thirty years is merely a small wave in the long river of history, but in the history of harnessing the Yellow River in the new China this has been a major achievement. These were thirty years of developing and exploring the river, thirty years of providing benefits and eliminating disasters, and thirty years of harnessing the Yellow River in which one victory after another has been won. On this world-renowned river with its massive amounts of sediment, there was built a great dam of over 300 meters in height and over 700 meters in width that was unprecedented in history. Like a great wall of water, the dam cut off the roaring Yellow River and like a steel chain, it bound the neck of the river and turned it from a dragon that brought perpetual disasters to the people to one that now provides them with benefits and aids the task of socialist construction.[21]

The Three Gate Gorge and Three Gorges Dams

In November 1994, the Three Gorges dam on the Yangtze River was officially launched—thirty-seven years after the Three Gate Gorge dam

---

*The virtual war that broke out over control of these lands is vividly described in Leng Meng, "Huanghe dayimin" (The Massive Population Resettlement on the Yellow River), *Zhongguo zuojia* (Chinese Writers), (n.p., 1996), pp. 60–92. Zhang Yue, the central government inspector who helped ease the confrontation in Xi'an, is the head of the Three Gorges resettlement feasibility group in Beijing.

had been started. What are the similarities between the two dams?

Both structures are on the mainstreams of the most important rivers in China. Both are record-breaking, large-scale projects designed to control floods downstream. Both are said to provide "huge economic benefits" through electricity generation: Originally, it was claimed that the Three Gate Gorge dam would fulfill one-third of the country's electricity needs, and now it is said that the Three Gorges dam will provide one-tenth of the national supply. Both dams are supposed to facilitate navigation on their rivers. Both projects require submerging vast tracts of farmland and relocating large numbers of people: 410,000 in the case of the Three Gate Gorge and 1,200,000 in the case of the Three Gorges. The Three Gate Gorge dam destroyed 1000–year-old cultural relics and antiquities from the civilization of central China. The Three Gorges dam will destroy relics from an even earlier period, the Ba Culture.

In both cases, provinces in the upper reaches of the two rivers suffered: Shaanxi in the Three Gate Gorge and Sichuan for the Three Gorges. Representatives from Sichuan pleaded to the government just like their counterparts in Shaanxi, but in both cases their opinions were systematically ignored. Both projects faced similar sediment problems. In the case of the Three Gate Gorge, sediment threatened the city of Xi'an while sediment trapped in the Three Gorges reservoir will threaten Chongqing. Both cities are provincial capitals and important industrial cities. Catchy slogans have been created to help deal with the dams' respective sediment problems—"soil conservation" was the clarion call of the Three Gate Gorge, just as "store clean water and flush out sediment" is the proposed solution for the Three Gorges.

Both projects chose to gradually increase the normal pool level and resettle people at different stages. The failures that beset the Three Gate Gorge in this regard will undoubtedly afflict the Three Gorges.

Both dams provoked intense debate, and in both cases minority opinions were censored and slandered. Both dams became the "concern" *(guanxin)* of the top leadership—Mao Zedong for the Three Gate Gorge and Deng Xiaoping for the Three Gorges. In each case, it was this "concern" that served as the primary force pushing the project forward.

Authorities claimed that both projects were "demanded by the people" because they were approved by the National People's Congress. And both were funded by the state as part of the planned economy. As a result, no matter how large the disaster, no one will have to take personal responsibility for their failures.

The potential social and political impacts of both projects were ignored. In each case, the head engineers (Wang Huayun for the Three Gate Gorge project and Lin Yishan for the Three Gorges) were the Party's "red specialists." Because of their red backgrounds, they feared nothing. Both were blindly confident and pursued the vanity of dictatorship.

In both cases, the dam's opponents—Wen Shanzhang in the case of the Three Gate Gorge and Huang Wanli in both cases—proposed smaller dams as the most rational alternative to large-scale dams and giant reservoirs. Huang's predictions for the Three Gate Gorge dam all came true. For the Three Gorges project he cautioned that it is absurd to build a dam on the mainstream of such a sediment-laden river. He also warned the authorities not to ignore local interests. Let's not forget that it was the Railroad Protection Movement in Sichuan Province that triggered the 1912 Republican revolution.

The Three Gate Gorge dam now has to be operated according to Wen Shanzhang's original instructions from 1957. What will happen in the case of the Three Gorges?

Huang Wanli, who has now passed away, will never see the Yangtze River blocked up. But what about Wen Shanzhang? Will he one day be forced to write the book on what went wrong with the Three Gorges dam project?

Notes

1. *Huanghe zhi* (Records of the Yellow River), Henan People's Publishing House.

2. "Sanmenxia shuili shuniu yunyong yanjiu wenji" (Compilation of Research Documents on the Operations of the Pivotal Three Gate Gorge Hydro Project) (Henan, Peoples' Publishing House, 1994).

3. Wang Tingji, "1956 Account of a Visit to the Soviet Union of the Engineering Bureau of the Three Gate Gorge Hydro Project," in *Zhongguo shuili fazhan shiliao* (Data on the Historical Development of Hydrology in China) (March 1989).

4. Cao Yingwang, *Zhou Enlai yu zhishui* (Zhou Enlai and the Harnessing of Rivers) (Department of Research on Party Literature, Central Committee of the CCP, 1991).

5. 1954 Report by Chinese Contingent, "Huanghe Sanmenxia shuili shuniu sheji jishu renwushu" (Technical Task Report on the Yellow River Three Gate Gorge Pivotal Hydro Project) (n.p., n.d.).

6. 1954 Report by Soviet Contingent, "Key Points of the Initial Design of the Three Gate Gorge Project" (n.p., n.d.).

7. Cao Yingwang, *Zhou Enlai yu zhishui*, 1991.

8. *Dui Zhengzhou Huanghe weiyuanhui gongcheng de caifang* (Interviews Concerning the Zhengzhou Yellow River Commission Project) (n.p., 1992).

9. Huang Wanli, *Zhishui yincao* (Harnessing the Water and Musings About Nature) (n.p., n.d.).

10. Huang Wanli, *Duiyu Sanmenxia shuiku xianxing guihua fangfa de yijian* (Views Concerning the Current Operational Plans of the Three Gate Gorge Dam Project) (n.p., n.d.). See also, "Huacong xiaoyu" (Small Discourses from a Flower Thicket), *Renmin ribao* (People's Daily), June 19, 1957.

11. Wen Shanzhang, *Dui Sanmenxia dianzhan de yijian* (Reflections on the Three Gate Gorge Power Station) (n.p., n.d.).

12. "Sanmenxia shuniu taolunhui" (Symposium on the Three Gate Gorge Pivotal Hydro Project), *Zhongguo shuili* (Chinese Hydrology), (1957–58).

13. Huang Wanli, *Duiyu Sanmenxia shuiku xianxiang guihua fangfa de yijian.*

14. "Sanmenxia shuniu taolunhui."

15. Opponents included: Wen Shanzhang, Huang Wanli, Ye Yongyi, Mei Changhua, Fang Zongdai, Zhang Shouyin, Wang Qianguang, Wang Tun, Yang Hongrun, Li Wenzhi, and others.

16. At about this time, a United States delegation had opposed construction of the dam at the Three Gate Gorge and had suggested Balihutong as an alternate site. Many Chinese hydrologists agreed with this suggestion but no one dared say anything in support of the American views. The Three Gate Gorge power station would be the largest in Asia and many comrades, out of patriotism, were anxious to launch this project. Also, no one dared disagree with the Soviet experts and support the Americans even if the issue was a purely technical one.

17. Wen Shanzhang, *Huanghe Sanmenxia gongcheng huigu yu pingjia* (Recollections and Evaluation of the Three Gate Gorge Dam Project on the Yellow River).

18. Wang Tingji, "Jin Zhou zongli zai Sanmenxia chaokaide yici huiyi," (Memories of Zhou Enlai at the Three Gate Gorge Conference), in *Zhongguo shuili fazhan shiliao* (Materials on the Development of Chinese Hydropower), No. 3, 1991.

19. Cao Yingwang, *Zhou Enlai yu zhishui.*

20. *Sanmenxia yimin shuiku nongcun yimin buchang biaozhun diaocha baogao* (Investigative Report on the Criteria for Reimbursement to Rural Refugees from the Three Gate Gorge Dam) (Internal Reference), Zhongguo shuili fadian gongcheng xuehui shuiku jingji zhuanye weiyuanhui (Reservoir Special Economic Committee of the Association of the China Hydropower Development Projects), 1990.

21. Ma Fuhai, "Zai jinian Sanmenxia shuili shuniu gongcheng jianshe sanshinian dahui shangde jianghua" (Speech at the meeting in commemoration of the construction of the Three Gate Gorge Project).

Chapter Eleven

Water Pollution in the Three Gorges Reservoir

Jin Hui, November 1993

Pollution of the Yangtze River

The Yangtze River has become the biggest sewer system in China. According to the Chinese Academy of Sciences, in the Three Gorges reservoir area alone "there are over 3,000 industrial and mining enterprises which release more than one billion tons of wastewater annually, containing more than 50 different pollutants."[*] By 1988, a total of 16.6 billion tons of wastewater had flowed into the Yangtze's basin from major industrial areas.[**] Included in the wastewater are such poisonous elements as mercury, cadmium, chromium, arsenic, phenol, lead, and cyanide.

Industrial sources account for the majority of the pollution, but significant amounts also come from agricultural run-off, residential wastewater, urban sewers, and pollution from ships. Most studies do not even consider the substantial amounts of pollution released from township-level enterprises.[†] Presently, there is very little treatment of industrial wastewater flowing into the reservoir area, and no treatment of residential wastewater;

---

[*] Yangtze River Water Resources Protection Science Research Institute and Environmental Evaluation Department, Chinese Academy of Sciences, *Changjiang sanxia shuili shuniu huanjing baogaoshu* (Environmental Report on the Pivotal Three Gorges Hydro Project on the Yangtze River) (n.p., n.d.).

[**]State Environmental Protection Bureau, *Huanjing tongji nianbao* (Environmental Statistics Yearbook) (n.p., 1988).

[†] According to Smil, the mushrooming of small and mid-size factories outside large cities does not bode well for China's water quality. "Industrial wastes released into the local lakes, rivers, and canals contaminate water that is used for drinking, animal care, and irrigation. No one knows with satisfactory accuracy how much unprocessed waste these rural enterprises release annually." Vaclav Smil, *Environmental Problems in China: Estimates of Economic Cost* (Honolulu, HI: East–West Center Special Reports, No. 5, April 1996), p. 25.

everyone relies on the river's capacity to flush pollutants out to sea to keep it clean. But following construction of the Three Gorges dam, the river's flow through the reservoir will be greatly reduced and with it the flushing capacity of the river.*

If we assume no net reduction or increase in wastewater release into the river, the COD level [chemical oxygen demand] in the water off the cities of Chongqing, Changshou, Fuling, and Wanxian will be greater by a factor of 2.08, 8.2, 10.36, and 3.5, respectively.** The overall water pollution index of the area contributing to the reservoir's pollution is presently 2.31, a level indicating moderate pollution. This figure will rise to 3.45 after the dam's construction, a level indicating a serious pollution problem. In Chongqing, for instance, the water pollution index will rise from 2.5 to 3.6 as a consequence of the dam's construction. In addition, the dam will create bays off the main reservoir at various sites. The slow moving and stagnant water in these reservoir bays will have a lower dissipation capacity in comparison to the mainstream, something that will further increase pollution problems. Moreover, prior to the first rainstorm of the year, there is likely to be a dearth or lack of oxygen in the water.

By the mid-1980s, a 500–kilometer-belt of pollution stretched along the Yangtze from Dukou, in Sichuan, to Chongqing, Wuhan, Nanjing, and Shanghai. In Chongqing, the white foam released from a nearby paper mill is now referred to by locals as "white ducks." The term was coined, oddly enough, by Prince Sihanouk of Cambodia. While on a tour of the Jialing River some twenty years ago, the then-Prince asked his guide what the white forms floating on the water in the distance were. His Chinese guide answered that they were flocks of white ducks! Twenty years later, the white ducks are still there, day and night.

I have traveled the entire length of the Three Gorges by ship, and have seen the sewer pipes located on the banks of the river. These pipes dump untreated waste into the river 24 hours a day, covering the water with pollution. The local boatmen hate these pipes with a passion and told me exactly how many there were and when each was installed. They told me

---

* Slow moving and stagnant polluted water in a reservoir creates the opportunity for massive algal blooms which deplete the oxygen from the deeper water. In some reservoirs, this oxygen-depleted water becomes acidic and toxic, killing all life in the river when released downstream.

**Chemical wastes discharged into a river can deplete oxygen levels, killing fish and other organisms. COD is a measure of this potential oxygen depletion.

River Pollution. *(Photo by Richard Hayman)*

how the number of pipes spilling pollution into the river has multiplied in recent years, how the mountains on both sides of the river have become increasingly barren, and how the water has become more and more polluted and the fish increasingly inedible. But they worry less about these problems than about the unknown diseases that have afflicted them and other workers on the river as a result, they feel, of drinking the river's water.*

---

*A study by the Chongqing Branch of the Yangtze River Transportation Bureau showed that more than 600 workers out of 2,200 suffered from various liver, stomach, and lung diseases.

According to the State Environmental Protection Bureau, in 1992, 58 percent of the Yangtze River and its tributaries had a water quality of Classes I and II—suitable for use as drinking water. Twenty-two percent was rated Class III—usable for drinking water after treatment. And 20 percent was at Classes IV and V—suitable only for industrial water supply, recreation, and irrigation.* In October 1993, the Yangtze River Water Environmental Monitoring Center issued its own report on the water quality of the Yangtze. It found that, "the water samples taken in September from around Fengjie, Wushan, Badong, Zigui, Sandouping, Nanjinguan, and Yichang were at Class III. The water quality in the area of the left bank of Yichang was at Class IV and that of some of the tributaries in the Three Gorges area were even worse—in some cases below Class V." These findings came in spite of a higher than average rate of flow in the river (and a consequent increase in the river's ability to flush pollutants out to sea) during the sampling period.

The monitoring center's report is clearly a warning. The water quality of the Yangtze River is deteriorating rapidly. Although the Supreme Creator endowed the Chinese people with a great river containing rich water resources, and despite ancient adages such as "the endless waters of the Yangtze pour forth" and "the ageless river is eternal," we must come to terms with the current reality: The river's natural capacity to purify itself is not infinite.

There are now more than 400 million people living in the Yangtze River Basin. An old song about the Yangtze proclaims: "You nurture the children of various nationalities with your sweet milk." These words now ring hollow because the sweet milk has become poison.

Lessons from Gezhouba

During construction of the Gezhouba dam, from 1970 to 1989, virtually no consideration was given to the potential effects of the reservoir on the water quality of the river. At that time, people were less aware of the impacts of dams on river ecology and the environment and, perhaps

---

*No source for the water classification system was cited in the original Chinese text. The World Bank, however, also uses a five-class system to rank water quality in China. See, for example, World Bank, "Staff Appraisal Report, Hubei Urban Environmental Project" (Washington, D.C.: The World Bank, 1995) (Rept. 14879–CHA), p. 47.

more importantly, they felt that the dam's reservoir was quite small
compared to the flow of the mighty Yangtze.* In 1981, as the dam
neared completion, the likelihood of a serious pollution problem
seemed remote. Instead, people were treated to the fantastic scenery of
a smooth lake emerging among the high gorges. The view from the
town of Yemingzhu, where the water was said to shine like a mirror,
was particularly attractive. There was talk of building a multipurpose
water park in the area, and the National Sports Committee even considered moving its water sports training facilities to the city of Yichang
(near Gezhouba) since its current site at East Lake on the outskirts of
Wuhan was seriously polluted.

But the environmental impacts of the dam's construction soon became
apparent.

Yemingzhu Port

On October 23, 1993, I visited the renowned port of Yemingzhu, and to
my great surprise I saw a river full of sewage, with garbage scattered
everywhere. The surface of the water was covered with oil and drifting
lotus plants. Moreover, the smell and color of the water were simply
unbearable.

According to an official from the Environmental Protection Bureau
(*Huanbaoju*) of Yichang, the pollution in the Gezhouba Reservoir
stems mainly from the following sources: waste, including oil, released
from ships lined up to pass through the dam's locks; seepage from
phosphorous (*lin*) deposits extracted from a local mine that have been
piled up on the riverbanks awaiting shipment; sewage released into the
reservoir from nearby residential areas and hospitals; and, finally, industrial wastewater.

A major polluter, the Number 403 factory, which produces ship en-
gines, releases waste oil into the reservoir via a network of small brooks.
When the accumulation of oil on the surface of the reservoir is particularly
heavy, nearby farmers skim off a few jars, pour it into their tractors, and
drive off. Fires also frequently break out on the reservoir when matches

---

*One likely reason why so little attention was paid to the possibility of increased
river pollution is that no detailed feasibility study of the project was ever conducted.
See, Shiu-hung Luk and Joseph Whitney, eds., *Megaproject: A Case Study of the
Three Gorges* (Armonk, N.Y.: M.E. Sharpe, Inc., 1993), p. 6.

are carelessly thrown into the water. As I was completing this study in November 1993, the water quality at Yemingzhu had deteriorated to Class IV, which is unsuitable for drinking. Nevertheless, 50,000 tons of drinking water is drawn daily into a local waterworks from the reservoir. Moreover, nitrate levels in the water have recently increased by 20 percent annually.*

The Huangbai and Xiaxita Tributaries

I also visited the banks of the Huangbai River, a primary tributary of the Yangtze, which flows through Yichang. Prior to the construction of the Gezhouba dam, the water quality of the Huangbai was very good. But when the dam was built and the reservoir backed water up into the Huangbai, it slowed down the flow of the river and consequently the ability of the Huangbai to flush out pollutants. Water quality in the river was recently lowered to Class III or IV, and an October 1993 study found that the water quality of the Xiaoxita tributary of the Huangbai River was rated at Class IV.** According to the local environmental protection officials with whom I spoke, the wawa fish, the Chinese Yangtze River sturgeon, and the rouge fish (yanzhi) have all disappeared from the area. Apparently, in 1991, entire truckloads of yellow phosphorous was dumped directly into the river, which caused the water to deteriorate dramatically and killed upward of 300,000 jin of fish. Fishermen who formerly raised fish in wooden cages submerged in the water were all driven away.

The renowned "Yangtze River China Sturgeon Research Institute," located on an island in the middle of the Xiaoxita tributary of the Huangbai River, can no longer draw water from the Huangbai for use in its sturgeon breeding efforts. The water is too polluted. Instead, researchers have to fetch water from the mainstream of the Yangtze miles upriver. Funding to

---

* Pollution of water with nitrates from high levels of nitrogen-based fertilizers is a growing problem in China. The well-documented health effects of high nitrate loadings include potentially life-threatening methemoglobinemia (also known as blue baby syndrome), which is characterized by a reduced ability of the blood to carry oxygen, and increased risks of stomach cancer. High nitrate loadings also contribute to eutrophication, the excessive algal blooms whose subsequent decay deprives affected waters of most of their oxygen supply and kills fish and other vertebrates. See, Smil, *Environmental Problems*, p. 25.

***Changjiang sanxia shui huanjing tongbao* (General Report on Water Quality in the Three Gorges Area of the Yangtze River) (n.p., n.d.).

Entrance to Small Three Gorges, Daning River *(Photo by Audrey Topping)*

transport large volumes of clean water is due to run out in 1994, leaving the future of the sturgeon uncertain. As for the poor residents of Yichang, they have no choice but to use the polluted waters for drinking and other domestic needs.

New studies are underway to evaluate the overall water quality of those portions of the Huangbai River affected by the Gezhouba dam. Hopefully, these studies will not only help save what's left of the Huangbai, but will also serve as an example of what may happen to the Yangtze River, as well as the Jialing and Wu rivers, and the Daning River's Small Three Gorges. People have only just begun to appreciate the immense beauty of

Small Three Gorges. *(Photo by Audrey Topping)*

the Small Three Gorges, but they too will soon be submerged by the Three Gorges dam. Right now, the pure waters of the Small Three Gorges are just about the only place left that leads one to appreciate the old adage about how "the waters of the Shu River [the name given to the Yangtze in Sichuan Province] are clear enough to reflect the Shu mountains" (*shushan shuibi shujiangqing*).

Who Pays for the Pollution?

At noon on October 22, 1993, some colleagues and I landed on Zhongbao Island, one of the construction sites of the Three Gorges dam. The island is actually no longer an island because the 813–metre-long coffer dam (built to sever and divert the river so that the main dam can be built) has connected it with the northern bank of the river. The officials in charge of the project from Hubei Province told me that construction began last year [December 1992], and that there are now over 5,000 workers and 800 pieces of heavy machinery working at the site day and night. He told us that China Central Television would carry a report on the first phase of construction work on the dam that very evening. However, because the National People's Congress (NPC) had not yet approved the final construction schedule for the dam [and would not do so until March 1994], dated pictures of the site would be used in the television report in order to create the impression that work on the dam had only just begun. The official explained that they did not want to be accused of looking down on or swindling the NPC, and asked that I not leak any information on the early start of the dam. It would only make it more difficult to maintain the NPC's support, he claimed.

So far in 1993 the state has invested ¥2 billion in the project, and in 1994 the figure will rise to ¥4.5 billion. As the rate of investment gathers steam, the completion date for the project has also been moved up. As a result, pollution control in the Three Gorges area is an even more urgent issue than I had imagined. The environmental impact statement for the dam states that "as much as ¥2.8 billion will be needed to clean up wastewater flowing into the river; a figure that is currently beyond the economic capacity of the region." It continued: "The volume of pollutants being dumped into the river must be reduced after the construction of the dam to compensate for the reduction in the rate of pollution clearance [flushing] from the reservoir. Input of pollutants to the reservoir should be reduced by an average of 43 percent of the current influx. Given the current volume of pollutants being discharged, 650,000 fewer tons of wastewater must be released each day." However, for every 650,000–ton reduction, ¥650 million of capital investment will be required, a figure not included in the original budget for the project.

The official cost estimate for the Three Gorges dam has surpassed ¥100 billion. How can we afford to invest billions of yuan to build this megaproject, when we cannot afford the ¥2.8 billion to treat wastewater in the reservoir area or the ¥650 million to mitigate pollution made worse by the dam. If we can afford to "buy the horse but not the saddle" we will embarrass ourselves in this unprecedented project.

And the cost of pollution abatement is by no means the only problem.

According to officials from various organizations in the Wanxian municipal government, the agricultural land surrounding Wanxian will only support 10 or 15 percent of the current population after the dam's
construction. The other 85 percent of Wanxian residents will have to be
resettled elsewhere. The officials also said that their research indicates
widespread opposition to resettlement if the move does not lead to an
improvement in living standards. If relocation causes a decline in living
standards, the locals will categorically refuse to move.

The essence of the resettlement policy for the rural or agricultural
population is to move people back from the river and resettle them in
nearby mountainous areas. But the slope of the land in the mountains is
much too steep for agricultural production, and many consider the plan
unworkable. Those forced to move up into the mountains will suffer lowered standards of living, and the environmental impacts of the policy will
be devastating. The Three Gorges area is one of the most fragile areas in
the entire Yangtze River basin. Even if the project was implemented per-
fectly, resettlement up into the mountains would still cause increased soil
erosion and pollution. Officials are correct in emphasizing that resettlement could be disastrous if not handled properly.

The pollution caused by township-level enterprises is also a serious
problem. Not long ago, the Chongqing municipal government approved a
new policy which requires the number of township enterprises to double
annually and be developed "in an unconventional and Great Leap Forward
fashion." In the eyes of the dam's planners, these enterprises are cash cows
that will play a key role in employing relocatees, alleviating poverty, and
paying off various local officials who might otherwise oppose the project.
But these enterprises are now developing at a pace beyond anyone's control. In areas of Sichuan Province, obsolete facilities for making coke,
sulphur, and arsenic still exist. The enterprises are protected by local officials and are totally beyond the control of the environmental agencies. As I
noted earlier, the pollution they cause is not even included in national
statistics. Sadly, large numbers of resettlers are slated for employment in
industrial and township enterprises. This will undoubtedly lead to the further destruction of the Yangtze River Valley's already fragile environment.

Environmental problems with the dam are cause for concern at the
Three Gorges Environmental Protection Bureau, which was established
in 1976. To date, there has not been any proper environmental planning
for the project, and the environmental budget is being administered by
the Yangtze River Planning Commission (*Changwei*), an organization

that because of its commitment to large-scale projects along the river has generally shown little concern for pollution alleviation. The new district of Wuqiao in Wanxian is a good example. No environmental plan for the district exists, and only those facilities which can afford to move have been relocated. Then there is the case of the Eastern Sichuan Chemical Company, which has received ¥2 billion from the government for construction of a new plant that will employ 20,000 relocatees. But the plant will produce 60,000 tons of caustic soda, 100,000 tons of formaldehyde, and 100,000 tons of polyvinyl chloride annually. Clearly, the construction of this plant will only add to the pollution problem in the reservoir, not solve it.

Conclusion

The Three Gorges dam will exacerbate an already serious pollution problem in the Yangtze River. By severing the mighty river and slowing the flow of its water, the dam will cause pollution from industrial, residential, and township-level sources to concentrate in the river rather than be flushed out to sea. The result, for the 400 million Chinese who live in the Yangtze River Basin, will be a poisoned river.

Chapter Twelve

Military Perspectives on the Three Gorges Project

Da Bing

There have always been strong links between key economic projects and national defense. This is the case with the Three Gorges dam. Dam supporters argue that the project will increase the supply of electricity and promote economic development, thereby strengthening the country's defenses. But because the dam will be built in central China, it will assume immense strategic importance. If it were destroyed by military attack, the consequences for the military, and for the entire nation, would be disastrous.

Although the world is becoming a more peaceful place, the threat of an attack on the Three Gorges dam cannot be dismissed.* Large cities, nuclear power plants, and hydro projects are routinely considered prime military targets. Examples abound: The British bombed Germany's Mohne and Eder dams during World War II, the United States bombed North Korean dams during the Korean War and Vietnamese dams and dikes during the Vietnam War, and in 1938 Chiang Kai-shek ordered his troops to blow up the Yellow River dikes to stem the Japanese Army's advance.

Based on available materials, it appears that the Leading Group for the Assessment of the Three Gorges Project (which issued its report in 1987) did consider how to protect the dam itself; however, if the dam were to come under attack it would most likely be part of broader multitarget assault. The enemy would not only target the Three Gorges dam but would also seek to destroy all the surrounding facilities, including the Gezhouba dam and the Zhi River Railway bridge. Therefore, any assessment of the

---

*China's perspectives on current international relations are examined in Robert Ross, ed., *East Asia in Transition: Toward a New Regional Order* (Armonk, N.Y.: M.E. Sharpe, Inc. 1995).

military threat to the Three Gorges dam must not be limited to an evaluation of the threat posed by a single strike against the dam, but instead must
include comprehensive defense planning for the entire region, which has
not yet been done.

China's defense system is weak and unable to defend against attacks by
modern weapons.*

Based on assessments so far (such as a twenty-year-old Chinese Air
Force report on the poor defensibility of the dam which, at the time, led
the People's Liberation Army [PLA] to oppose the project),** our primary
defense is to lower the water level in the reservoir in a crisis situation in
order to reduce the threat of disastrous floods in the lower reaches if the
dam were to be destroyed. While we might receive advance warning of an
attack, it is by no means certain. Therefore, strengthening and improving
our defenses must be seriously considered.

The effects of a military attack on the Three Gorges dam would be
serious and widespread. But most experts limit their analysis to the impacts of a single attack on the dam and the floods that would result. And
indeed, the floods caused by even a single strike would be disastrous—the
entire area from the dam site at Sandouping to just above Shashi City
would be flooded by a wall of water rushing forth at 400,000 cubic yards
per second. A minimum 130,000 to 208,000 people on the southern bank
of the river would be affected. However, the limited, single strike analysis
concludes that the dam burst would not seriously affect the Jin River
dikes, Shashi City, and the area around Wuhan. Its impact, they conclude,
would be primarily local.

But as we mentioned earlier, any complete assessment of risk should
focus not just on the destruction of the dam and resulting floods, but also
on other potential impacts; for instance, the destruction of shipping routes,
the loss of electricity generation, and possible contamination from nuclear
or biological weapons. A military attack on the Three Gorges dam would
clearly not just have local impacts.

If, along with the Three Gorges dam, the Gezhouba dam, the Jin River

---

* For a recent assessment of China's military capabilities, see *New York Times,*
December 3, 1996.

**Yang Lang, "High Dam: The Sword of Damocles," in Dai Qing, *Yangtze!
Yangtze!,* pp. 229–241, and Winchester, *The River at the Center of the World,* p.
236. The current involvement of many PLA-run companies in the dam's construction
may have helped swing elements of the military to support the project.

dikes, and the Zhi River bridge were also attacked, the destructive impact would increase exponentially. In addition to the local flooding mentioned above, the flood waters would reach the Han River plains, destroying railway lines and directly threatening the Wuhan area. Tens of millions of people and over 300,000 square kilometers in the plains would be threatened. Railways, roads, and navigation routes (especially the Beijing-Canton and Jiao-Liu rail lines), would be destroyed, making it very difficult to move supplies and personnel north or south.* The navigation route between Wuhan and Chongqing would also be destroyed, creating similar difficulties for the movement of troops and supplies between the east and west. The resulting immobility of troops would seriously compromise China's ability to sustain front-line forces in any coastal conflicts. The destruction of the Three Gorges and Gezhouba dams would also cause blackouts in several provinces in central and east China and would halt military movements and production in western Hubei, eastern Sichuan, and western Hunan. Finally, the use of nuclear or biochemical weapons would create immense amounts of contamination and pollution throughout central China.

During World War II, British attacks on dams in the Ruhr Valley had a paralysing effect on German industry and were particularly devastating to the German war effort. The U.S. bombing of dams during the Korean War had the same effect: In summer 1952, the United States bombed four power stations—the Fushan, Jiashan, Changjing, and Shuifeng power stations, which together comprised 13 separate power plants with a total capacity of 1.524 million kilowatts.** Of the 13 plants, 12 were completely destroyed, causing a two-week blackout throughout North Korea and a 50 percent reduction in industrial production in and around Pyongyang and the Yalu River. The impact of the bombing was also felt in northeast China where, even with the supplementary electricity provided by a series of small-scale dams and power stations, electrical supply was cut by 90 percent. Later, in June 1953, another five North Korean dams were bombed by the United States. The damage severed the water supply to 75 percent of North Korea's rice paddies, seriously disrupted

---

*The major Yangtze River flood in 1954 cut off the Beijing-Canton line for 100 days, causing widespread industrial and commercial havoc.

**Power plants in Baghdad were also destroyed by Allied forces in the recent Persian Gulf war.

transportation, and caused enormous psychological stress amongst the population. An earth and rockfill dam 32 kilometers north of Pyongyang was also bombed (flooding 43 kilometers of land and cutting major railway and transportation routes) as were the Guicheng and Deshang dams which were situated near key transportation routes into northeast China.

The severity of the destruction caused by a military strike against a dam is not directly proportional to the size of the dam and reservoir; it also depends on the location of the facility and the extent of preventive measures taken before an attack. The Soviet Union, France, Austria, and the United States have all conducted studies and taken various measures to mitigate the effects of military strikes against their key hydro installations.

But the advent of nuclear weapons and precision-guided cruise missiles have made it increasingly difficult to guard against attack. Many analysts have therefore argued that any plan to construct large-scale reservoirs must include, as a priority, the consideration of military risk. Why then did the experts who carried out the assessment of the Three Gorges dam pay so little attention to this issue? And, how could they believe that as long as a warning system was available for releasing water in the midst of a crisis, losses could be minimized? What if the warnings were inaccurate? And what effects would releasing water have on electricity generation, especially during a time of war? Would navigation by 10,000–ton vessels still be possible with a lower reservoir? And if the enemy were to destroy the Three Gorges and Gezhouba dams, the Jin River dikes, and the Zhi River bridge all at once, what would the total damage be?

Since the real purpose behind the assessment of the Three Gorges dam was to promote an early launch to the project, it is obvious that the experts simply decided to avoid the issue of war altogether since the questions mentioned here are very difficult to answer. The Yangtze Valley Planning Office's assurances that "appropriate engineering and managerial measures would limit the flood damage caused by conventional or nuclear attack" result not from an honest assessment of this problem, but from a concerted effort to sweep the problem under the rug.

The assessment's cursory review was not the last word on the issue, however. An article entitled "War and Its Impact on the Three Gorges," published in *Qiyejia* (Entrepreneur) magazine, raised the eyebrows of many and finally brought the issue to the forefront. According to the article:

> Yin Zhishu, a defense expert from the PLA General Staff, has recalled that as early as 1959 Premier Zhou Enlai placed Zhang Aiping, then deputy

director of the General Staff and chairman of the National Defense Science Commission, in charge of tests of air defense for the proposed Three Gorges project. Zhang Aiping, Qian Zhengying, Lin Yishan, and 60 other scientists and experts from various military, scientific, and hydro-planning organizations were gathered together to carry out the evaluation.

Yin Zhishu claims that in the course of these tests, explosions and simulated direct attacks were conducted. The findings recommended lowering the water level during a crisis, diverting water into the lower reaches, reinforcing the coffer constructions in front of the dam, and building air defense facilities. To deal with a direct attack on the dam, the tests called for the design of a large-scale concrete structure and the piling of rocks. . . . After the research was completed, General Zhang delivered his report to Premier Zhou and later it was passed on to experts during the 1987 Assessment.

During the 1960s, the engineer corps, the Yangtze River Planning Office, and the Hydro Research Institute participated in nuclear weapons testing in China's remote Xinjiang Province. More than ten tests were conducted. . . . The Hydro Research Institute conducted seven tests on four dams, including air and land attack scenarios and the emptying of the reservoir. The Electric Research Institute focused on testing the effects of war on electricity transmission and electrical equipment.

Because of the size of the Three Gorges dam, it has a certain capacity to resist heavy bombs. Zhang Aiping proposed that studies on the possible destruction of the dam by atomic and hydrogen bombs be made a priority. The basic air defense policy was to walk on both legs, that is, to adopt active and passive air defenses at the same time so as to be able to eliminate the enemy before it reaches the target and . . . to ensure that the destruction which does occur is minimized.

Despite once again raising the issue of a military defense of the Three Gorges, the article was filled with errors and angered General Zhang Aiping when he read it.

The true account, in Zhang Aiping's own words, is as follows: "In August 1958, I followed Premier Zhou's order and, along with Comrade Zhang Zhengman, went to Yichang to study the air defense system for the proposed Three Gorges dam at Sandouping. The results of our aerial surveys of the river and surrounding lands showed that defense against conventional aerial attack would be no problem. Studies by the Yangtze Valley Planning Office came to the same conclusion. But a nuclear attack was something completely different. If the dam were bombed in a nuclear attack, the water would rush all the way down to the city of Nanjing, which would suffer enormous damage, as would the Dongting and Poyang lakes. The only defense would be to release water from the reservoir ahead of time, but this is extremely risky, for there is no way to know when the enemy will attack. Moreover, the Three Gorges dam will bring no benefits to Sichuan Province let alone to the greater southwestern part of China. And in the upper reaches the dam will cause Wanxian County to be submerged and threaten the area around Chongqing. The dam will not benefit navigation and will cause sedimentation to accumulate in the upper reaches of the Yangtze and increase the likelihood of flooding.

"We reported all of this to Premier Zhou. Zhou asked my personal opinion about the project and I said frankly, 'Let future generations decide this issue.'

"The article claims that 'As early as 1959 Premier Zhou Enlai placed Zhang Aiping . . . in charge of tests of air defense for the proposed Three Gorges project.' " But, in reality, the tests only used hand grenades and small bombs. The nuclear tests in Xinjiang were also limited and didn't really achieve the expected results.

"All in all, Three Gorges defense research after 1959 focused primarily on defense against nuclear attack. But since it comes so swiftly, and because we could never predict such an attack, we concentrated on passive defense methods such as reinforcing the dam base and releasing water before an attack. But we were not sure of the measures' effectiveness. As for the notion that we can predict when war will break out, I don't know what to say about that!"

As Zhang makes clear, there are no easy solutions to the problem of defending the Three Gorges dam, and research into the problem has been woefully insufficient. The 1987 Assessment dealt with the problem by simply assuming that no war would ever threaten the dam and unleash the destructive powers of the water which lies behind its wall.

The New Golden Triangle of China

Richard Hayman

"Don't you understand? We can't guarantee your safety!" barked the gravel-voiced Public Security officer. The radio was crackling from headquarters: "How did they get there? They can't stay! Send them down!" I translated for my two friends who were having their first encounter with the rude officiousness of the Chinese police.

It was November 1995, and we had come to hike into the highland region known as Shennongjia, home to many prized rare plants and animals, including the legendary *ye ren* ("big foot") primate. These mountains, on the borders of Sichuan and Hubei provinces, rise to over 10,000 feet in sheer cliffs and steep ridges over rushing tributaries of the Yangtze River in the Three Gorges region. The cliffs are riddled with ancient caves of unknown depth, many of which will be flooded by the reservoir. We had met herbalists in the town of Dachang on the Daning River and examined many odd roots and fungi. Our guide, Old Zhu, offered to take us up into the highlands and introduce us to collectors. Little did he know how much trouble he would get us into.

We had been riding in the back of a local truck eating dust and diesel fumes for six hours. The steep mountains rolled past, climbing ever higher. The slopes are deforested and studded with gray boulders. Mud-walled houses nest in bamboo groves. Farmers hoe sweet potatoes and harvest chili peppers. At one turn in the road, majestic cedars stood above a government office. "They're only for scenery. The rest are gone," noted a truckmate. We got off the truck in Guanyang, a town of mud buildings with one street paved with large flagstones. There was an unusual hubbub on our arrival. A crowd of wide-eyed mountain people stared at us and examined our packs. The driver was worried because the police were calling over to us. They had never had foreigners here before.

"Dragon" caves along the river likely to be flooded. *(Photo by Audrey Topping)*

Our protests were for naught. "For our safety" we would leave as soon as a military jeep was summoned. I could not persuade the grim-faced police and Party officials that we would assume responsibility for our own safety while hiking, as we often had in other counties. "There are no bandits, but no, you can't go in. It's decided," the cop grunted.

Later, out on the street, I noticed a large new government center with bright tiled walls that was under construction. It seemed like an odd building for such an out of the way hill station. The crowd of onlookers was pressing on us again, so I asked them about the herb collecting in the area. They started giggling and muttering quietly, so I asked the question again. Finally, a youth blurted out: "There's lots of opium!" Some of the others

Caves along the river to be flooded; unknown where water will flow.
(Photo by Richard Hayman)

shouted at him to be quiet and then slipped away, but the cat was out of the bag. A local official then came over, shooed everyone away, and ushered us into a jeep. Soon we were bouncing away in a stream of dust on our way down to the lowlands.

Back in Dachang, we reported to the police station. They wanted us to leave immediately for the county seat, but the only means of transportation were the river craft that depart at dawn every day and travel through the gorges. Unable to leave, we returned to the small inn where we had stayed earlier, and wandered around the ancient village. Dachang has many exquisite buildings that date from the 1500s, a few of which are scheduled to be dismantled for reconstruction in a new, higher town when the Three Gorges dam floods the valley. We talked with a few townsfolk and farmers who confirmed that opium is indeed being grown in the area. They explained that opium had become a booming export business, carefully guarded by the police and military.

The area had been a major opium production area from the mid-1800s until the rise of the Chinese Communist Party (CCP) in 1949. The drug was a source of cash for the local warlords and a competitor of Indian opium brought in by the British and Americans. The "Wanxian Incident" of 1927, when British gunboats bombarded Wanxian on the banks of the Yangtze, was precipitated by the struggle to control the opium trade. Now it appeared that the cultivation of opium is enjoying a resurgence in southwest China. "The Communists used to suppress the trade, now they do it themselves," said one local farmer.

Conversations about the Three Gorges dam led to similar expressions of dismay. "How are we going to live? The resettlement land up above is all rocks!" Billboards in the hills mark the elevation to let everyone see how high the reservoir will rise. Above, in the steep mountains, work crews use explosives to blast new terraces for the resettlement program. We hiked up to see the new fields. A farmer was piling rocks for terrace borders, but the soil was mostly broken rock. "How will you grow anything in this?" I asked. "I will bring soil and fertilizer up the mountain," he answered with a forlorn look. Another more outspoken farmer chimed in: "We will not survive, but we can't do anything about it." "Can you protest?" I asked. "How?" was the reply. "That is just asking for trouble. The nation says to move. We will see what happens when the waters come."

The solution for many farmers forced to move up to the marginal lands may be to grow opium. The reality is that it may be the only cash crop other than tobacco that can support the ever-growing rural population. Official plans call for industrial development to provide jobs for displaced peasants, but the poor state of roads, low education levels, and lack of resources belie these plans. Meanwhile, the opium trade grows.

I was told by locals that the state tobacco companies have their hands in the processing, and that the police and military successfully transport the opium through official checkpoints. Opium paste and refined heroin is sent by truck to Chongqing, Chengdu, and Kunming and then flown to Guangzhou for international shipment. Ever more of the drug is spilling out onto the streets of China, used by bored underemployed youth. It keeps them passive and ready for exploitation, and it is evident in labor mills and widespread prostitution.

The involvement in the opium trade of local officials out for personal gain is a poorly kept secret. Higher officials certainly know about it and have attempted to impose some controls. But this region, as the old saying goes, "is far from the emperor." The extra cash may also be needed to make up for local shortfalls in funding for the resettlement costs of the Three Gorges dam. It may sound farfetched to suggest that opium profits would be used to defray social costs, but similar situations have occurred in Colombia where some drug money has been used to build local projects and to reward "cooperative" communities. As long as foreigners are kept out (as we were) and official inspectors are well treated, there is no stopping this easy money.

Later, we were escorted out of Dachang and down river to the county seat of Wushan which crowds the banks of the muddy Yangtze at the entrance to the Wu Gorge. A police detachment met us at the dock with its sirens blaring and led us through the crowded streets to the police head-quarters for a long day of questioning. We were handled fairly politely because of my position with a shipping company that brings much business to Wushan. I had my friends protest loudly in English while I bargained apologetically in Chinese. The police insisted on inspecting our packs, but allowed me to conduct the search. I unpacked our camping gear onto their desks and made as much of a mess possible. Then I opened the container which held the gas for our camp stove and prepared to pour it out for inspection. Anxious policemen quickly extinguished their cigarettes. With the first two sleeping bags rumpled in a disorderly fashion across the desks, I was about to open the third stuff-sack when the police said they did not need to check it. There were about 40 rolls of exposed film hidden in the bottom of that third sack.

We were then asked to hand over our film and notebooks. We had prepared fake notes and rolled back blank film for just such an occasion. When they insisted on developing the film, we went to a local photo shop that could not develop slides. I had put in one exposed roll of scenery which I suggested might come out. It was all green and vague, but seemed to satisfy the police chief's curiosity. The other rolls would be sent out to be developed later in Chongqing.

I was interviewed for a formal transcription about our breach of security—illegal research, talking to locals, photographing government installations (bags of herbs at collecting stations), and unauthorized travel. I made an obsequious apology assuming all responsibility and appealed that Old Zhu and the local people we talked to not be punished. "Good attitude!" beamed the police chief.

After signing a confession and a receipt for the confiscated film, notes, and an old mushroom found in my bag, we were released on the condition that we take the first ship out. Opium was never mentioned.

Progress on the dam

As 1996 comes to a close, the Three Gorges dam project in China's Yangtze River Valley is developing rapidly. The construction is surprising in both scope and pace. And it is now evident that the region is undergoing a dramatic and reckless transformation as a result of the dam.

I recently spent three months traveling in the dam and reservoir area, working as an advisor for a new tourism project on the Yangtze. A fleet of six new river cruisers is being launched to carry the growing number of international visitors rushing to see this unique and endangered area. Whatever one's opinion of the Three Gorges dam project, the world is coming to witness the spectacle of its construction. It must be seen to be believed. The world's most magnificent riverscape is rapidly becoming its most tragic.

Dam construction is well underway. As a priority project in the current Five-Year Plan, over U.S.$1 billion a year is being poured into the project. The summer of 1996 saw the completion of the coffer dam along the south bank of the river. A large diversion abutment is ready at the head of the coffer dam, which will divert the full flood flow of the river—over 100,000 cubic feet per second—around the natural riverbed while the main dam is being built. It is projected that the current in the diversion channel will be so strong as to require massive tug boats to tow much of the river shipping around the dam site. Long traffic delays are expected.

The navigation authorities have given notice that a rise of a few meters in the mean river levels may occur after the diversion. The effects may not extend beyond Zigui—75 kilometers upstream of the dam site—but it has been suggested that a higher water level may force reluctant residents to move before the year 2003, when the reservoir is scheduled to rise. If this is true, the coffer dam, which now stands about 15 meters above the average river flow, would need to be raised.

The north bank has deep cuts in the hills where the lock systems are being built. The five-step flush locks and one-tank shiplift already have their entrance channels prepared. People's Liberation Army demolition teams have dynamited out the hills on the riverbank to form the locks.* Massive earthmoving equipment is kept busy carving out the banks of the river, while gravel works lay substrate. A deep pit has been dug at river-edge on the north bank to prepare the foundation for the massive dam.

Rising above each bank, at the center of the site, is a series of giant placards with large red characters which read: "*Yi liu guanli, yi liu zhiliang, yi liu shigong, wenming jianshe,*" or "First Class Management, High-Quality Workmanship, First-Rate Construction."** On the south bank even larger placards proclaim: "*Kaifa sanxia, fazhan Changjiang,*" or "Build the Three Gorges, Develop the Yangtze."

Just downstream of the dam site, a graceful suspension bridge over one kilometer long has been completed by the Wuhan Bridge Construction Company. The bridge has already alleviated the long lines of traffic that used to wait for ferries. Along the north bank, the highway from Yichang is mostly complete, bringing ever more traffic to the booming metropolis. The steep canyons along the river have been filled with rubble from the road construction. Nearby cement plants are in full operation, providing cement facing along the river banks. A new dock on the north bank allows visitors to disembark at the new city center where the project management, research, and training offices are located.

In the Xiling Gorge at Xintan, cliffs that overhang the river channel are being reinforced by a mining engineering company. These towering rocks are likely to be undermined by the reservoir waters and collapse. Should the rocks give way and slide into the reservoir, massive waves would surge forth and cause significant destruction.† A landslide in 1984 on the north

---

*Construction gangs overseen by the Hydropower Control Department of the People's Armed Police are also involved in the project, along with an unknown number of prison laborers. *Wen Wei Po* (Hong Kong), May 28, 1996.

**Problems encountered in the course of construction and the importance of project quality control have already become the source of concern by some Chinese leaders who are intimately familiar with the perennial problems of engineering in China's hydroelectric sector, such as the Gezhouba dam which has already shown severe signs of structural deterioration despite its recent completion in 1989. See, statement by Qiao Shi, Xinhua, June 19, 1996.

†Not according to Chinese officials, however, who claim that even under the "worst-case scenario [of a landslide], the waves would raise the reservoir level by 2.7 meters at the dam site, posing no threat to the dam or other structures." Xinhua, June 16, 1996.

Bridge construction over the Yangtze at Three Gorges site. *(Photo by Richard Hayman)*

bank destroyed half of a local village and closed the river for months.

No section of the gorges is immune from the ravages of development. Only the south bank below Huangling Temple has been left relatively untouched, with a few farms still growing oranges. Even below the dam, in the lower Xiling Gorge (which will not be inundated), the landscape is being mauled for massive quarries to gather rocks for the widespread construction.

Nanjin Pass, at the entrance to Xiling Gorge, is home to numerous resort villas which are being built into the hillsides to accommodate visiting officials and business representatives. A new cable car across the river above historic Three Travelers' Cave is now in operation to provide views of Xiling Gorge and the broad Gezhouba dam just below.

As the pace of destruction accelerates, so too does the tourism industry in hopes of cashing in on the "natural wonders" to be lost to the reservoir. Travel agencies around the world advertise "last chance" tours and cruises, falsely claiming the reservoir will flood the region in 1997. The advertising is working—the number of visitors to the Three Gorges area is growing by 25 percent per year. And cruises through the region (from Chongqing downstream to Shashi or Wuhan) are now a standard feature of many tours of China.

There are now more than fifty river cruisers offering various grades of service plying the Yangtze's waters in the Three Gorges region. Among the many new ships recently launched is the *Three Kingdoms*, designed like a third-century warship complete with flags, bells, faux cannon, and oars painted on the hull. Another new ship, the *Qianlong,* is in the shape of a gigantic dragon: bridge officers navigate through the teeth of the dragon's head and a golden tail rises aft. The hull is built in early Qing dynasty (1644–1911) palace style, with numerous plastic ornamentations and scenery viewing pavilions. These two ships are owned by the Pinghu Company of Yichang, which is partly owned by the Gezhouba Dam Authority, which reinvests profits from the dam.

The most expensive and elegantly designed of the new ships is the *Shenzhou,* or Sunshine. It has large cabins and a two-tiered dance floor, private karaoke rooms, and a Japanese garden complete with a flowing stream. The Sunshine was financed in part by the State Council for use by national leaders and their guests. The dining room features a large mural of the completed dam, in all its misty glory.

Meanwhile, outside of all the picture windows, the life and death of the Great River unfolds. On any given day, bodies can be seen bobbing in the Yangtze's rapids, the tragic victims of treacherous waters in a land where few people learn to swim. These *shui da bang,* or "waterlogs" in Sichuanese, are rarely retrieved from the water, eventually becoming "water ghosts."

The summer of 1996 saw record floods in the middle reaches of the Yangtze. Most of the flood waters flowed from the Han and Xiang rivers downstream of the dam site. The Three Gorges dam would not have stopped the floods. The levees on the Jianghan Plain in Hubei Province had to be reinforced to hold back the surging waters which reached 28 meters above mean low at Wuhan, a historic high. Near Dongting Lake in Hunan Province, numerous villages were washed away, and 10,000 people were drowned. Much of the flooded land was part of flood diversion areas that had been filled in or reclaimed in recent decades; it eroded away swiftly. Fishermen I spoke with in Yueyang lamented the loss of life but showed me their bounty of large fish. "This year the fish had plenty to eat!" they said.

Upstream from the dam site, another new bridge is under construction near Wanxian. A high arch bridge, it is reputedly the largest of its design in the world. The footings are set into the cliffs at a narrows in the river and seem to be set so low that they may soon be under the reservoir water level. The bridge will certainly be a boon to this region which will be hardest hit by resettlement. New terraces are being cut high in the mountains to accommodate farmers, and lowland soil is being carried up in bamboo backpacks to add to the rocks. Many farmers have resigned themselves to the fact that farming will be difficult, if not impossible, in the mountains. They will accept a one-time resettlement buyout and try to find new ways to make a living. This is the critical test of the social impact of the dam project.

Politically, the most important recent change was the designation of Chongqing as a national municipality on par with Beijing, Shanghai, and Tianjin, with its authority extending to the border of Hubei Province. The official population of the city-region is about 14 million, making it the largest municipality in China. Many expect Chongqing to become the Shanghai of southwestern China, to drive the development of the interior, and to complement and keep up with the more prosperous coastal areas.

As a political entity independent of Sichuan Province, Chongqing now has considerable latitude to make reforms and arrangements according to "practicalities." The local papers trumpet *gaige kaifang*—reform and opening up—as a great opportunity. The underlying reality is that the region must reinvent itself economically to survive the engulfing of the busy towns and productive land.

The city is in dramatic transformation. The waterfronts along the Yangtze and Jialing rivers are being rebuilt with highways and new docks. The old shanty houses and warehouses that used to store goods carried off ships and up the muddy riverbanks by coolies are now gone. In their place, a "Great Wall" of concrete has risen: an embankment 30 meters high with tree-lined roads. Now, steps lead down to gangways over the rushing current to floating docks where the ships tie up. Most passengers now dock here and hike up the stairs upon disembarking where the *bang bang jun* (the "help, help army") of coolies vie for porterage work. The landing is unfinished and dangerous, and is especially slippery in the rain. The waterfront is scheduled for completion in spring 1997, but locals say the funds have been embezzled away.

As part of Chongqing's designation as a national entity, its entire political leadership is being replaced by appointees from Beijing. Local officials are being fired under the banner of "reform" and against a background of profligate corruption. The Party secretary has resigned after revelations that his son was living the high life in Hong Kong, driving a Rolls Royce and engaging in numerous shady business deals. The deputy mayor, Wei Wenli, is in custody and under investigation after losing U.S.$60 million in foreign exchange trading by the city-owned Yu Feng Company. Another deputy mayor is in jail after allegedly embezzling U.S.$2 million in foreign trade dealings.

The suppression of local opposition to the dam may be another reason for the firing of the local leadership and the new federal status of Chongqing and the gorges region. More and more complaints over resettlement and funding are being heard as the realities of the dam project are being felt. Local officials who speak the local language will not be as resolute as outside authorities. Moreover, the vast sums of resettlement money provide great opportunities for corruption. Already, numerous officials have been cited for abuses.

Meanwhile, construction continues everywhere. At Wushan, a new hotel, called the Three Gorges Guest House, has the typical marble and mirror glitz of modern China. When I stayed there, the new but shabby rooms had no hot water in the shower. When I asked the manager for hot water he directed me to the new massage parlor for a "full bath." In downtown Wushan a new billboard details the resettlement schedule. Thousands will be moved each year, even though many facilities are not yet ready. The entire town will be moved before the year 2000.

The town of Zigui is being moved to a new site on the south bank of the river near the dam. New apartment towers can already be seen from the river. The Qu Yuan Temple relocation has also begun, and a new museum is being built to house the stone stele inscriptions of Qu Yuan's classic poetry. But the most complete relocation program is at Badong, where the new town, just above the old, is largely finished. The construction on the steep hills has caused landslides which destroyed a road and factory at the water's edge.

Rural relocation has also been tested in Badong County, the most underdeveloped region of the gorges. The native Tujia minority people have been moved up into the mountains, where there is little arable land, increasing soil erosion, and little hope of sustenance. Local officials are seeking industrial investment to provide jobs, but the educational level of

Population resettlement schedule in Wuhan. *(Photo by Richard Hayman)*

the population is so low and the resources and transportation so lacking that there is little hope. As a result, some officials have turned to opium, which is smuggled out to the south for export as a cash crop.

The new leadership in Chongqing is assuming authority over a city in the midst of wrenching transformations. Entire neighborhoods are being torn down, their residents driven to the countryside. Residents are promised comparable new apartments, and many are eager to move to better conditions with plumbing and electricity, but most are not relodged for years, while expensive housing sits empty waiting for the nouveau riche. The California Gardens in the Jiangbei section on the north side of Chongqing comprises a dozen 30–storey towers, new malls, and a giant bowling alley. Fanciful "villas" with neon-lit sculptures sit across the valley near the new expressway to the new airport. Nearby, a surviving Song dynasty (twelfth century) ceremonial gate to a lost temple is cramped in neglect between construction rubble and a gas station. The graffiti I saw scrawled on one condemned house sums up the views of many who are forced to move: "No money to go. No place to go."

Massive office and shopping complexes are rising in the center of the city. One, Twenty-first Century Plaza, has a seven-story atrium that features a lovely waterfall and fountain surrounded by a bamboo garden. Thousands of peasants from the countryside visit the glittering skyline city every day, and more than a few availed themselves of the Plaza waters for practical uses and relieved themselves in it until the fountain was so fragrant that it had to be shut down.

Over 100 million gallons of pollution are estimated to be dumped into the Yangtze every year, much of it industrial toxins. Paper, steel, silk, and chemical factories line the river in Sichuan, often covering the swirling waters with foam or effluents. Some particularly odious pollution sources are being closed, but little money is being directed to refit factories or enforce pollution controls. Recent examinations by China's Environmental Protection Agency have shown "Class IV" serious pollution at hot spots in the gorges region. The accumulated waste of many towns is simply piled by the river and washed away during flood season. There are laws against all this, but no enforcement. The Yangtze has always been used as a grand flush. Uncontrolled, the discharges will doom the water quality of the reservoir, making the waters virtually useless for irrigation or fishing. One of the official reasons for building the dam is the future diversion of river waters through an aqueduct to irrigate the ever drier north China plain in the next century.* The water delivered may be so polluted as to be useless.

Work to salvage the archaeological heritage of the Three Gorges area is as chronically underfunded as is pollution abatement. A Three Gorges Museum was opened in Wanxian two years ago, but it displays only a Ba people hanging coffin and a few photos of Han dynasty *que* towers from Zhongxian County.

Financial and political pressures are mounting around the dam project. Competition for funds within the government is leading to increased opposition to the project from other government departments. A series of foreign bond offers have been canceled or postponed since 1995 for lack of interest. A recently offered $120 million in bonds will "test the market" once again. Hundreds of foreign firms are competing to win contracts for the massive dam, especially where their investments are guaranteed by public funds. Key to the confidence of the international community are the "stability" of the government, and an open examination of the dam's designs—neither assured for now.

This "Great Wall" across the Yangtze is certain to raise the stakes in the ongoing power intrigues in Beijing. The dam is the most divisive issue

---

*The South-to-North Water Diversion Project will reportedly employ nuclear explosions to cut canals through rugged mountains.

in China, becoming a "Great Leap Forward" on the Yangtze bound for rough rapids. A recent poll taken among CCP cadres before Deng Xiaoping's death in February 1997 found that most expected "turmoil" after his death.

The dam project may come to a halt with any major political reshuffle. Once Premier Li Peng finishes his current term of office in March 1998—and if he is not appointed to any other position—the dam project may be "reconsidered" and scaled back to a more manageable size without its extreme social and environmental costs. A Japanese government delegation of hydroelectric experts and officials was received in October by Li Peng and took one of my company's cruise ships to Yichang to inspect the dam site. In an aside to me, one of the Japanese delegates said that the group doubted the claims of the dam officials and did not plan to participate in the project. They felt the dam as planned would not survive Li Peng's exit.*

Even in that event, a series of low rise dams would likely be built, as recommended by most international consulting agencies. Then the world would certainly rush in to help finish the job, if the river dragon ever allows the waters to be stopped.

---

*In December 1996, Japan's Export-Import bank offered credit to Japanese companies bidding on part of the Three Gorges project. The Swiss, German, and Swedish governments have also indicated their intention to provide similar public financing.

Acknowledgments from "General Plan for Population Resettlement"

Li Boning

This volume, *Developmental Resettlement Is Good,* was compiled at the suggestion of Comrade Cheng Zihua before he passed away.* In 1986, as a member of the Three Gorges Project Coordinating Group under the Central Committee, Comrade Cheng took an extensive inspection tour of the Three Gorges area. He met with and heard the reports of cadres at different levels of government from Sichuan, Guangxi, and various other counties and municipalities and investigated local resettlement trial projects extensively. He visited the experimental rice paddy plot in Zigui County and watched a video titled "The Advantages of Developmental Population Resettlement." He gave a very positive review of the experiences gained from the trial projects and subsequently suggested to me that since the Three Gorges project was not well-understood in Chinese society and there was even considerable misunderstanding of it, a volume be compiled describing the good results produced by the trial projects. This, he felt, would help people understand the real situation and ease their worries and concerns. Based on this suggestion, we compiled the first volume of this work and have plans to publish one or two additional volumes.

This volume reflects the experiences of the five-year trial projects. The data and articles contained in it draw from typical experiences in county and municipal trial projects, assessment reports, articles about resettlement prepared by district and municipal leaders, and extensive discussions about relevant issues raised by the trial projects. Although the trial pro-

---

*This appendix was originally included with Li Boning's section of Chapter Four.

jects are still at the preliminary or experimental stage, they point us in the right direction for implementing resettlement and provide us with valuable experiences to help deal with the problems created by the process.

In this volume there are more articles on rural resettlement than on urban resettlement, and with good reason—rural resettlement is one of the most difficult issues we face and it is the number one priority of a successful resettlement policy. Included herein is an article on The First Team of Leijiaping Village, Dongrangkou Township titled, "Explore Land Resources Within the Team and the Initial Success of Moving the Entire Village Back from the River and Resettling in Nearby Areas." Hongyan Village, Renhe Township, Fengdu County, provided another article titled "Experiences in Planting Orchards for Relocatees Using Available Local Resources and Moving Back from the River and Resettling in Nearby Areas." Shuitianba Township, Zigui County, concentrated on the development of local land resources, and its valuable experience in moving back and resettling in nearby areas can be applied in other villages. Xingshan County explored a number of measures that are described in the article titled "Gaining Benefits Before Inundation and Good Planning After Inundation." Baoping Township, Yunyang County, expanded its existing orchards. Wuling Township, Wanxian County, expanded its town-owned orchards which years ago had been a sent-down educated youth farm (*zhishi qingnian nongchan*) and provided experiences in township-run orchards. Shizhu County provided an article titled "Develop Land to Resettle Relocatees and Increase Profits." Changshou County reviewed the experience of developing a local fruit known as sandy tangerines and of laying a foundation for resettlement. Badong County provided experiences in using resources on state-owned farms to plant orchards for relocatees. Taipingxi Township, Yichang County, provided experience from developing tea farms on which relocatees can be resettled. Fengjie County provided experiences in animal husbandry and similar endeavors.

The experiences drawn from these trial projects in rural population resettlement are varied and rich and reflect the success of these projects. They also indicate that there is sufficient space for resettling people. This not only guarantees that the people's current living standards will not decline, but indicates that they will prosper and may get rich.

Though there are fewer articles on urban resettlement, the volume provides general information on its success. It also points out that urban resettlement needs to focus on the construction of road, water, and electricity projects and must control construction below the submersion line. Another important focus is education, which is touched on in many articles and which is key to the success of resettlement work. Why is the orange technology program organized by the Badong Population Resettlement Bureau so welcome? And why did the same program in Zhenxi Township of Fuling Municipality achieve such rapid results? The key to their success was the considerable attention paid to education.

Resettlement is an important part of and a critical factor in the success of the entire Three Gorges project. Our five-year experience demonstrates that we must reform the lump-sum reimbursement practice followed in past resettlement projects. Each case in this volume examines one aspect of the new method and points out where improvements should be made.

The purpose of this book is, on the one hand, to inform society and help people from all walks of life understand the situation and, on the other hand, to play a role in exchanging experiences in the Three Gorges reservoir area so as to generate support for the project and do a good job in resettlement work. For this reason, we sincerely hope that those who are concerned with and those who participated in trial projects contribute good articles for future publications.

I would like to extend my gratitude to Vice Premier Zou Jiahua, who wrote the inscription on the book flyleaf, to State Council member Chen Junsheng, to Vice President Wang Renzhong, who wrote the calligraphy on the cover of the book, and to all those comrades who expressed an interest in the book. We would also like to extend our gratitude for the concern and support we received from comrades in the central government, which will undoubtedly inspire cadres and masses in the Three Gorges area to devote even greater efforts toward the construction of the project. It is a pity that our beloved Comrade Cheng Zihua passed away before the book's publication. But on his sickbed his concern for the project remained and he too wrote inscriptions for the flyleaf. Let us dedicate this book as an everlasting memorial to this great proletarian revolutionary as well as to our revolutionary predecessors.

Appendix B

Sediment Problems at the Three Gorges Dam

Luna B. Leopold, April 1996

Editors' Note

This piece was prepared by Professor Leopold in April 1996 as part of a submission to the United States Export-Import Bank. At that time, the Ex-Im Bank was considering whether to provide financing and loan guarantees for U.S. firms bidding on contracts to either supply equipment for the Three Gorges dam or help build it. Citing the requirements of its environmental guidelines, the Ex-Im Bank subsequently announced that it would not provide such assistance for U.S. firms.

The Three Gorges dam is designed to operate under conditions practically untested in the world and never before tested in such a large structure. Projections of controlling sedimentation within the reservoir are subject to significant uncertainties. China has about 83,000 reservoirs built for various purposes, of which 330 are major in size. Sediment deposition in 230 of them has become a significant problem, resulting in a combined loss of 14 percent of the total storage capacity. In some, more than 50 percent of the storage capacity has been lost.[1]

The proposed operating procedure at Three Gorges is as follows: During the flood season (May through September), the reservoir level would be held at a low pool elevation, called the flood control level (FCL). During this period inflows are used for power generation. After the flood season, the lower flows with lower sediment concentration will be impounded [behind the dam], and the reservoir pool level will rise to the normal pool level (NPL).

It is proposed that in time, sediment will be deposited in the reservoir until a uniform slope of the bed will be established that can just transport the annual sediment load through the reservoir. It is estimated that this condition will be reached between 70 and 150 years after construction, depending on the choice of the elevation for the FCL.

There are an estimated 17 reservoirs in the world that operate in this manner, seven of which are in China and one in the United States.[2] All of these, except one, are small in size. That exception is Sanmenxia in China, which has only 18 percent the storage capacity of Three Gorges—in other words, very much smaller. Thus the world's experience with this type of operation is very meager.

These uncertainties lead to the conclusion that forecasts that presume to support the financial benefits of the dam may be in error and that investing in the project would be unwise. There are several sources of such error in dredging costs, resettlement costs, flood control benefits, and other areas, but the present discussion deals with the problems only in the subject of sedimentation.

The sedimentation conditions at various times during the first 100 years of the dam's operation have been forecast by use of mathematical models and physical analogues that involve many assumptions of unverified reliability. Any difference between the forecast and on-the-ground performance has large financial, environmental, and human implications. Therefore, it is necessary to specify some of the most important possible sources of error in the forecasts.

The largest dam construction projects in the United States—Hoover, Glen Canyon, Bonneville, Fort Peck, Tennessee Valley, to name a few— do not utilize the principle of sediment flushing or sluicing contemplated for the Three Gorges, but some of the same uncertainties were present in their design. The American experience gives some direct insight into possible future uncertainties. Perhaps the most important lesson gleaned from direct experience is that conditions fifty years in the future are usually quite different from any forecast, and one hundred years in the future are simply not forecastable.

One problem is in the ability of designers to forecast the rate of sediment accumulation in a reservoir. Even when the records of sediment inflow are reliable, the deposit rate is often quite unanticipated. For the multipurpose reservoirs in India, Murty states that the "annual loss rates of siltation in most reservoirs are 145 percent to 875 percent of the figures assumed at the time of construction."[3]

The Canadian Yangtze Joint Venture (CYJV) estimates that the equilibrium would be reached when the 90 to 95 percent of the sediment entering is flushed through the reservoir.* This would be in about 100 years.[4] A forecast of an approximate condition 100 years ahead is hardly a financial surety.

A most important problem involves the actual management of the facility. Between May 1 and September 30 each year, large sediment-laden flows of water will be discharged in order to carry away sediment. But those same large flows of water cannot be used to fill the reservoir to provide for the winter needs. Moreover, the large flows that carry the sediment can also be the cause of floods. Because the high sediment inflow corresponds closely with the high water inflow, the needs of flood storage and sediment removal are antithetical.

In the case of a possible flood condition upstream of the Three Gorges, the prudent course of action would be to close or partially close the discharge gates so that potentially destructive floods would be prevented by storage of the incoming water. But the incoming high flood flow also carries the most sediment and could not be flushed through the reservoir. The sediment so held as a deposit in the reservoir settles on the bed and requires more force to dislodge it than was necessary to keep it flowing to the outlet gates.

Year to year, this simultaneous need to pass sediment through the reservoir and the need to store water for power or flood control requires a neat and sophisticated day-to-day forecast of inflow of water and sediment. The hope is that a slight lack of simultaneity of water and sediment inflows could be used to move sediment just ahead of or behind the greatest flow of water.

Experience in dam operation in the United States shows that such delicate management is uncommon. The delicacy of such an operation is emphasized by the fact that the sediment rating curves do not show the common lack of coincidence of maximum water flow and maximum sediment flow. Usually the sediment is greatest in the early or rising limb of the annual hydrograph, but at the Three Gorges, the rating curves are not

*The CYJV study was a Cdn$14 million feasibility report funded by the Canadian International Development Agency and completed by four Canadian engineering firms: Hydro Quebec, SNC-Lavalin, Acres International, and BC Hydro. The study, now widely discredited, concluded that a 175–meter dam was feasible. Although the Chinese are now building a 185–meter-high dam, detailed Chinese studies of sedimentation are not widely available. For a critique of the CYJV study, see Margaret Barber and Gráinne Ryder, eds. *Damming the Three Gorges: What Dam-Builders Don't Want You to Know,* 2nd ed. (Toronto: Earthscan Canada, 1993).

looped.[5] Therefore, dependence on a lack of coincidence is a poor procedure on which simultaneous flood control and sediment flushing must rest.

Another source of possible trouble is in the forecast, 90 to 100 years in the future, of the final slope of the deposited sediment in the reservoir. If the slope is greater than forecast, the deposition in the channel at the head of the reservoir would be much steeper than forecast, leading to unanticipated flooding. The analyses of this important matter were done by computation and models, with no detailed analyses of experience in other reservoirs of the world. Only one page in the CYJV report was devoted to another example—that of Sanmenxia reservoir in China—but no statement was made concerning the depositional slope. This one example is hardly reassuring because after two years of operation, 1960-62, the sediment deposition was so large that the operation of the reservoir was completely altered. "New tunnels were driven . . . and some penstocks were converted to spillways."[6] The rate of deposition had been grossly underestimated, and remedial action was needed after only two years.

Another possible problem is in the assumptions regarding bedload, the coarse gravel and sand that will eventually accumulate at the head of the reservoir. The gravel component in the sediment was deemed so small that it was "not considered in reservoir sedimentation calculations."[7] However, the gravel portion is of great importance at the head of the reservoir. The report suggests that "gravel bedload amounting to perhaps 200,000 cubic meters per year may need to be dredged for the Chongqing reach" for the chosen elevation of the FCL.

The bulk of the sediment that will be deposited in the reservoir will be the sand portion. The gravel- and cobble-size material will be the first to drop out of the flow and will accumulate near the head of the reservoir. The slope of this coarse deposit will determine how far upstream it will extend and thus the extent to which it will result in flooding nearby Chongqing and harm the navigation channel and facilities. The project plan apparently expects that as the final condition is reached in 100 years, all the incoming gravel will have to be dredged, each year, into the indefinite future. If the incoming load is underestimated, these costs could be so financially burdensome that the original benefit-cost relationship will be quite discredited.

The effect of sediment storage in the reservoir on the channels downstream is given little importance in the study. The report implies that until equilibrium is established, about 100 years hence, water with little sediment will be discharged and "there will be degradation below Gezhouba and the alluvial reaches of the middle and lower Yangtze."[8] But the report says "some degradation may be beneficial."

Experience in the United States of degradation by clear water below a dam hardly justifies such optimism. Below Hoover Dam on the Colorado River, the degradation was some 35 feet. Below Fort Peck on the Missouri, there was serious bank erosion. Discharge of water with a low sediment content for 100 years is not likely to be insignificant in its effect on downstream channels. If degradation is great, diversion works could be destroyed. If bank erosion is serious, the extensive levees may be in jeopardy—levees critical to the flood management system of the river's lower reaches.

Downstream of the Three Gorges, the alluvial plain is settled by several millions of people who all depend on diversion works for irrigation water and on massive levies to confine floodwaters. The morphology and stability of the channels on this alluvial plain are conditioned by the combination of water and sediment that has characterized the river for hundreds of years. If clear water from Three Gorges flows into such a channel part of each year for many decades, the channel will react.[9] Experience in many countries demonstrates that the reaction will be some combination of bed erosion and bank erosion.

Downcutting could leave diversion works and canals high above the river level and thus require new engineering facilities to correct the problem. Bank erosion would tend to undermine the flood control levees and thus demand levee rehabilitation.

The analysis of the CYJV says that the effect of sediment on the discharge tubes in the dam would be minimal. It speaks only of abrasion to the penstocks and turbines. But the experience in the United States is that the discharge of large volumes of water through tunnels, pipes, and penstocks results in serious cavitation, due to local below-atmospheric pressure resulting in pieces of rock and concrete blown off the walls of the tunnels, especially at the entrance. This creates a maintenance problem as well as the need to refrain from extended discharge at high rates for long periods. This potential problem deserves more attention than has been given in the analyses at Three Gorges.

Notes

1. Hu Chunhong, "Controlling Reservoir Sedimentation in China," *Hydropower and Dams* (March 1995): 50-52.

2. G.L. Morris and P.R. Rao, "Workshop on Management of Reservoir Sedimentation" (1991).

3. K.S. Murty, "Soil Erosion in India, River Sedimentation," *International Res. and Train Center on Erosion,* vol. 1 (1989).

4. Canadian Yangtze Joint Venture, "Three Gorges Water Control Project Feasibility Study," 5 (1988): 1–4.

5. Canadian Yangtze Joint Venture, "Three Gorges Water Control Project Feasibility Study," 5 (1988): 5–19.

6. Canadian Yangtze Joint Venture, "Three Gorges Water Control Project Feasibility Study," 5 (1988): 7–20.

7. Canadian Yangtze Joint Venture, "Three Gorges Water Control Project Feasibility Study," 5 (1988): 1–2.

8. Canadian Yangtze Joint Venture, "Three Gorges Water Control Project Feasibility Study," 5 (1988): 1–10.

9. China Yangtze Three Gorges Development Corporation, *Environmental Impact Statement for the Yangtze Three Gorges* (n.p., n.d.).

The Three Gorges Dam and the Fate of China's Southern Heritage

Elizabeth Childs-Johnson and Lawrence R. Sullivan

The fate of archaeology in the Three Gorges area is perilous, if not fatal, due to the intention to flood the middle Yangtze River within a mere ten years.* Archaeologists in China have been forced to scramble to preserve what they can of the past without provision by the government for adequate funding, tools, or time. Of the nearly 1,300 known sites along the 482 square kilometers of river bank, archaeologists have determined that between 400 and 500 are worthy of preservation. They estimate, however, that it will be possible to preserve only half of them.

Sites to be inundated by the Yangtze flooding comprise two major types: those primarily architectural monuments standing above ground, and those that are subsurface and require excavation. Finds dating from before the Han period (206 B.C.–A.D. 220) from the Three Gorges area are almost entirely archaeological in origin. Han and post-Han finds are represented by a select group of extant monuments, in the form of architecture and of engraved stones from the river's framing walls. The extent to which this area was originally dominated by a rich array of monuments has long been known historically. The Three Gorges is the alleged setting for the romantic poet Qu Yuan (338–278 B.C.), traditionally recognized as the author of the *Chuzi* (*Songs of the South*), and for the bloody feuds of the Three Kingdoms (c. A.D. 220–280) heroes, who were later celebrated in the Ming period (1368–1644) novel *Sanguo Yanyi (Romance of the Three Kingdoms)*. In characterizing this environmental melting pot of per-

---

*An earlier version of this article was published in *Orientations* (Hong Kong) (July/August 1996): 55–61.

Qu Yuan burial site. (*Zhongguo Changjiang Sanxia*, Hong Kong, 1993, p. 250.)

sonalities, author Lyman P. Van Slyke lyrically refers to Qu Yuan as one who "spoke in a characteristically 'southern' idiom: extravagant language, passionate emotional tone, a richly inhabited spirit world complete with seductive shaman goddesses, transparent sensuality, lush images of plants as metaphors—orchids for purity, fragrant flowers for virtue, weeds for evil."* Qu Yuan, the "protean archetype of personal integrity, loyalty and dissent" is celebrated not only in southern China's annual Dragon Boat Festival, but by a memorial temple in the Three Gorges at the alleged site of his burial, Zigui. The Qu Yuan temple was built during the Qing period (1644–1911) and restored during the 1980s. The government intends to transfer this entire modern building to a site removed from the flooding waters of the Yangtze, while the alleged burial site will be inundated.

The Three Kingdoms hero General Zhang Fei, who at the famous Peach Garden in 220 took an oath of loyalty in the face of death, is also celebrated by a temple built in his honor further up the river at Yunyang in Sichuan Province. The Zhang Fei temple was built during the Northern Song (960–1126) and was restored late in the Qing period by the Tongzhi emperor (r. 1862–73). Characterized by extensive gardens, the multiroom

---

*Lyman P. Van Slyke, *Yangtze: Nature, History and the River* (Stanford: Stanford University Press, 1988), p. 136.

complex with balconies sits at the foot of a mountain, where it precipitously overlooks the southern banks of the Yangtze opposite Yunyang Town. Zhang Fei was said to have been murdered here when his army officers mutinied. His sworn friend Liu Bei, with the help of the adept strategist Zhu Geliang, defeated the arrogant challenger to the throne, Cao Cao, in a climatic battle nearby at Red Cliffs, just to the southeast of the Three Gorges, between Hanzhou (Wuhan, Hubei Province) and Dongting Lake in Hunan Province. The battle sealed the fate of the Han and divided China into three kingdoms, and the site is one of the most famous along the Yangtze River due to its being celebrated in Su Dongpo's (1037–1101) time-honored prose-poem "The Red Cliff," and in numerous Song and Yuan (1279–1368) period paintings.

Yu Weichao, director of the National History Museum of China and an archaeologist currently in charge of excavations in the Three Gorges, has referred to the probable demolition of other equally remarkable architectural monuments, including the famous Shibaozhai of Ming period date—a Buddhist temple complex used until the Cultural Revoluation and located west of the Three Gorges at Zhongxian in Sichuan Province. This so-called fortified "treasury" leading up to the mountain outcrop of the temple site was built out of wood by the Jiaqing emperor (r. 1796–1820) in 1800 into the limestone rock face. At 56 meters, it is the highest building of its kind in China. The twelve-story edifice resembles a pagoda and houses Buddhist sculptures and slab stelae, one of which is inscribed with the history of the founding of the site.

Also in danger of inundation is the town of Dachang in Sichuan Province, which preserves Ming and early Qing folk architecture. Located on the eastern bank of the Daning River, the town was surveyed in 1957–58 by the Sichuan Museum, and in 1993 the Sichuan Institute of Archaeology carried out a preliminary excavation. Apparently, since the site's location was pivotal for north-south boat transport on the Daning River, its history may be traced archaeologically back to the Three Kingdoms or Jin (265–420) period. The special characteristics of this town's later history are the architectural design of a courtyard conjoining four houses and the enclosing volcanic wall with segregated views. Bold and uninhibited modelling of the latter wall's edge gives the site a strong and lively local flavor. Although the town of three main streets rises 141 meters above sea level, it will be completely inundated and has been ranked as a national-level monument. There are plans to move the entire town to drier, higher ground.

One other type of extant monument, unique to the middle Yangtze and its gorges, are the examples of "low water calligraphy" (*kushuitike*) engraved into critical passages along the Yangtze's limestone walls, beginning at Chongqing, but most numerous in the Three Gorges area. These engraved examples of historical and calligraphic value range in date from the Han and Eastern Jin (317-420) through the Qing periods. The major purpose of these so-called "low water calligraphies" is as reference points for recording low-water and thus safe or unsafe passage levels for cargo and boats. The most artistically prominent example, recently declared a national-level monument, are the Baiheliang engravings. Amounting to 163 sections and 30,000 characters, the Baiheliang engravings are located near Fuling in Sichuan Province and range in date from the Tang dynasty (618–906) to the Qing period, with the most numerous dating to the Song dynasty. The earliest Tang record encloses two profile fish, known as *shiyu* (stone fish), whose eyes when exposed were allegedly the symbol for a safe water level. An inscription of 917 explains:

> In the spring, February, of the Tang Guangde era [763–64] the river waters receded and the stone fish appeared so that four *chi* [Chinese feet] were visible below the fish. According to legend, the [seer] Xian said: When the Yangtze waters recede [and] the stone fish appear, the year will witness a rich harvest.*

Other still-standing architectural monuments include the Han period *que* (towers) represented by the Wuming (No Name) tower and Dingfang temple towers in Zhongxian. Like other *que* elsewhere in Sichuan Province, the Wuming towers of Ganjinggou and the pair from Dingfang temple at Eastern Gate of Zhongxian typify the pairs of tall, elaborately carved stone pillars traditionally positioned at the entrance to *shendao* (spirit ways) that in turn may preface large earthen burial mounds. Like other examples from the Chengdu area, these are stone replications of wooden prototypes, 3 to 4.5 meters tall, with a two-level bracket-type capital and tiled roofs. The upper surfaces of these towers are often elaborately decorated with cosmological themes in relief. Animals symbolizing the four directions, including the dragon of the east, tiger of the west, bird of the south, and tortoise and snake of the north, decorate the pillar bases.

---

*Zhuan Xiutang et al., "Sanxia gongcheng kuchu shuiwen wenwu zongshu" (Summary of Calligraphical Antiquities in the Three Gorges Reservoir Area) in *Changjiang wenhua lunji* (Symposium Papers on the Cultures of the Yangtze River) (Wuhan: Hubei Education Publishers, 1995), p. 20.

Semihuman dwarfs uphold corbeled eaves and various immortals and mythological creatures playfully decorate in deeply undercut relief the spaces between simulated beams and curving bracket ends. Immortals ride deer, play *liubo* (chess), or chase fantastic animals.

Buried remains from the Three Gorges area are as dramatic in quality and significance to the cultural history of China as the architectural and aboveground monuments of Han through Qing date. Their fate is equally perilous. The current archaeological data available from reports on the Three Gorges area provide preliminary documentation on what appears to be a significant but little understood southern, Yangtze cradle of civilization. Meaningful finds range in date from the Upper Paleolithic (50,000–12,000 B.C.) through the Warring States period of the Eastern Zhou (481–221 B.C.). The discovery of remains of *Homo sapiens sapiens* man at Ziyang and Tongliang between Chongqing and Chengdu and the more recent discovery of the even earlier *Homo erectus* in the Three Gorges area reveal typological similarities to modern Mongoloid populations and suggest that, with more archaeological evidence, it may be possible to qualify how this southwestern existence of early man contributed to the evolution of human prehistory in early China and the rest of Asia.

Before the Three Gorges area became part of the Chinese empire under the Qin (221–206 B.C.) and Han dynasties, the area was inhabited by a culture and peoples known historically as the Ba. It is hypothesized that the Ba may have originated in the advanced Neolithic cultures called Daxi (c. ?5000–3200 B.C.) and the succeeding Chujialing (c. 3200–2300 B.C.), which flourished in the area encompassing both banks of the Yangtze in the area of Dachang, and which has been excavated by members of the History Department of Sichuan University. What is going on west of this site, however, appears to be connected with another still unknown Neolithic culture, which hopefully can be explored through future, government-supported excavation. Distinctive wares of the Daxi and Chujialing phases include those polished red on the outside and black on the inside in the form of tall *gu*-shaped drinking goblets and those *dou*-shaped bowls placed on openwork stands. The flare for eccentric shape and abstract decoration differentiates these wares from the much more mundane painted pots of the northern Neolithic cultures of the Yellow River basin. Fine wheel-thrown, egg-shell thin blackwares and jades with shamanic imagery characterize the successive Shijiahe (c. 2500–2000 B.C.) and Erlitou (c. 2100–1800 B.C.) cultural phases, the latter of which has been identified at a site in Xiling Gorge (Zhongbao Island, Yichang, western Hubei Province).

Unfortunately, Zhongbao Island has now been completely inundated in preparation for the dam. The remains from this site were published in 1987 and belong to three levels that included the Daxi, Chujialing, and "late Erlitou-like" periods.* The "Erlitou-like" settlement is identified by the typical bag-legged pitcher called *gui*. When it is with a pipe-spout it is called *he*.

It is important to put into perspective that these Neolithic cultures of eastern Sichuan Province share with their western Sichuan neighbor, Shu, certain southern cultural idiosyncrasies. One of the latter is the motif of an elegantly shaped bird head with long curving beak that forms the handle of ceramic ladles. This same bird, interestingly, continues in the imagery of the bronzes of the succeeding early Shang period (1766?-1122? B.C.).

Although the Erlitou, Shang, and Western Zhou (c.1127?-771 B.C.) periods are difficult to document in the Three Gorges area and middle Yangtze basin, certain data suggest that this era was a creatively active one. Yu Weichao, among others, has reported that the region from Wuhan below Yichang west to Yunyang, including especially the Wushan area and the Xiling Gorge, is rich in Ba cultural data dating to the late Neolithic and Shang periods. Particularly exciting are the preliminary finds from Shuangyantang, located in the Daning River Valley just northwest of Dachang. The Institute of Archaeology in Beijing began preliminary investigations in 1983 and uncovered an area of 10,000 square meters. It is not yet clear if Erlitou remains from this site are comparable to those identified with contemporaneous remains elsewhere at Bai Miao, Lujiahe, and Yidu in the Three Gorges.** Early and middle Shang period finds are now represented by the recent discovery of a ritual bronze wine vessel (*zun*) that is of an approximate height of 80 centimeters. The shape and decoration of this vessel are not only identical to others found in western Sichuan at Sanxingdui and at Funanxian in Anhui Province, but also to another from Chengguxian, Shaanxi. Seated birds, which are unique to these early southern bronzes, alternate with the more familiar animal forms on the shoulders of these *zun*. This bronze discovery is exciting new evidence that points to the existence of what must have been a thriving,

---

Kaogu xuebao (Journal of Archaeology) 1 (1987): 45–98.

**Chen Xianyi et al., "Lun Hebei diqu cao Shang wenhua" (On the Early Shang Culture in Hebei), *Changjiang wenhua lunji* (Essays on Yangtze Valley Cultures) (Wuhan: Hubei Education Publishers, 1995), pp. 149–150.

independent southern bronze-working culture stretching across all of Sichuan Province and southern China during the Shang period.

Equally exciting are the excavations of another early culture called Lijiaba, near Gaoyang Town and Qingshu Village west of Three Gorges at Yunyang. A fourth season of excavations was carried out by the Archaeological Institute of Sichuan in 1995, at which time the site was designated "national level." The site is estimated as measuring 20,000 square meters in size at a depth of 1.5 to 2 meters. Lijiaba cultural finds are also represented at Ganjinggou in Zhongxian and Lujiahe in Hubei Province. All conspicuously hug low terraces of river valleys, extending from hills into valley floors along the river or river branches. At Lijiaba, lower-level finds date to the Shang and Spring and Autumn (722–481 B.C.) periods, and upper-level strata date to the Warring States (480–221 B.C.) and Han periods.

Although very little is known archaeologically of the Western Zhou period, it is known from historical records that the Ba were on good working terms with the Zhou ruling house, whose seat of power was to their north up the Han River at Xi'an in Shaanxi Province. Like the Shu of western Sichuan, the Ba helped the Zhou defeat the Shang; unlike the Shu, the Ba were apparently singled out for their loyalty to the Zhou King Wu (r.c. 1122-1115 B.C.).* The Zhou not only designated the Ba royal house with the surname Ji, but the Zhou took Ba women as wives. That there was cultural contact during the Shang period between the Zhou and Ba of the upper Han River and in the middle Yangtze regions further south is demonstrated, for example, by the shared bronze *zun* type with shoulder birds. It has been suggested that the Ba, as early as the late Shang period, were either in direct contact and living along the upper Han River as they were later in the Warring States period, or they were simply a major southern influence on the Zhou, who were gaining power in the area of the upper Han and further north.

There is no holistic view of the Ba from their alleged beginnings in Neolithic times through the Western Zhou period, or of the Ba during the Eastern Zhou period. Nonetheless, according to current assessments, distinctive Ba works of art of the Eastern Zhou/Warring States period include the *chunyu* (war drum) and *zheng* (gong), boat-shaped wooden coffins, a unique form of writing, the *ge* (dagger axe) decorated with tigers in profile, and other eccentrically shaped and decorated weapons. Except for the *chunyu,* the above

---

*Stephen Sage, *Ancient Sichuan and the Unification of China* (Albany, N.Y.: State University of New York Press, 1992), p. 53.

Example of "boat-shaped coffins" retrieved from Three Gorges area in a local museum. *(Photo by Audrey Topping)*

bronzes can also be found at Shu sites in western Sichuan Province, which has led to the archaeological term Ba-Shu to describe the period of circa the fifth to third century B.C., when cultural sharing between the Ba and Shu seems to have been at its height. The *chunyu* excavated from western Hubei Province and at Wanxian in Sichuan Province are typically bulbous in shape with a long cylindrical body and bulging shoulder and rim. Their lids, which can be removed, are usually crowned by a crouching feline, the ubiquitous Ba emblem. Although unknown at Shu sites, these drums are found as far afield as Guizhou, Yunnan, Hunan, and Anhui provinces.* Large drums reach 70 centime-

---

*Wenwu (Antiquities), no. 8 (1984), and no. 3 (1990).

ters in height and the smaller, portable ones average 20 centimeters in height. The latter were allegedly used in war chants and as a signaling device during war.* They were complemented by the *zheng,* which could produce a contrasting sound to the drum for long-distance communication. They have handles by which they can be held and struck. *Zheng* have been found at Xiaotianxi in Fuling, at Entuoshi in Hubei Province,** and at Shupu, Dajiangkou, in Hunan Province. *Zheng* can be large, measuring over 40 centimeters in height. The Ba aristocratic emblem of a tiger frequently appears cast on the outer surface of these instruments.

The function of the unique script of the Ba, which appears most commonly on bronze *mao* (spear points) and *lian* (short swords), is difficult to identify. Li Xueqin has argued that there are two forms of Ba script: "A" script is considered to be emblematic, but with both phonetic and ideographic components; "B" script is considered to be writing identical with Chinese.† The undeciphered BA script, including the hand, tiger, and bird motifs, are represented by Ba examples excavated at Xiaotianxi and Majiaxiang. The same graphs appear on *lian* excavated from a tomb in territory traditionally thought of as Chu, Jiangling in Hubei Province, which is just down the Yangtze River from Yichang. It is likely that these graph types found elsewhere on Ba-Shu and Chu weapons are signs of noble rank and ownership. Use of the graphs, however, gradually died out. Nor were any of the graphs integrated into scripts in use in northern China.

The Ba Culture of the Warring States period is characterized by a taste for eccentric weaponry. No doubt this disposition favoring weapons as objects of beauty distinguishes Ba expression. It is known from historical accounts that the Ba and Shu joined together to defeat the state of Chu on Ba's eastern flank during the early fourth century B.C. The revolver-shaped *ge* is allegedly Ba in origin, although it is better known in the much more thoroughly excavated Shu sites of western Sichuan Province, and when published is therefore labeled Ba-Shu. This distinctive *ge* type has a long shaft and wide blade simulating a revolver in shape and is usually decorated with a profile of a growling tiger. According to Li Xueqin, *ge* of this type are known from the sites of Fengjie and Fuling

---

*Sage, *Ancient Sichuan,* p. 55.

**Wenwu* (Antiquities) 3 (1980): 44.

†Li Xueqin, *Eastern Zhou and Qin Civilizations,* trans. Kwang-chih Chang (New Haven: Yale University Press, 1985), pp. 215–216.

Chu state culture site already excavated. *(Photo courtesy of Jim Williams)*

along the Three Gorges and middle Yangtze region, although most are unpublished.* The prototype of this weapon is the considerably earlier Shang *ge* made out of jade or bronze found at sites along the northern Yellow River. Other diagnostic Ba weapon types include the boot-shaped *yue* (axe), the three-cornered *ge*, and the frequently inscribed hollow *mao* and willow-leaf-shaped *lian*.

Whether or not the presence and influence of the Chu state and culture can be documented in the Three Gorges in part depends on the interpretation of the material character of finds from sites such as the Qin period tomb at Xiaotianxi in Fuling, allegedly the royal burial center of Ba nobility and kings. Although there is debate about the cultural identity of Chu deep in Ba territory, there are few features typical of Chu in this area during the Eastern Zhou period. In separate papers, Barry Blakeley of Seton Hall University and Wang Ran of Wuhan University demonstrate how through historical interpolation the area occupied by the Ba on the Three Gorges at Zigui,** for example, was confused with the capital of Chu at Danyang.

---

* Li Xueqin, *Eastern Zhou,* p. 208.
**See, Barry Blakeley, "In Search of Danyang. I: Historical Geography and

Wang also clarified why there are so few Chu remains in the Three Gorges area. Based on written records, it is stated that Ba and Chu were on good terms during the eighth century B.C. and that as late as 590 B.C. the Chu King Gong took a Ba woman in marriage. By the fourth century B.C., and after a great deal of warfare, Chu evidently penetrated Ba territory. What appears amid Warring States and early Qin and Han period remains in the Three Gorges is in fact primarily Ba; Ba never foregoes its cultural heritage of weapon and musical instrument types. From the one published tomb at Xiaotianxi there are numerous Ba artifacts, including round or boot-shaped *yue,* revolver-shaped *ge,* short-handed *zheng,* willow-leaf-shaped *lian,* but most importantly, *chunyu* and Ba writing. The inlaid bronze *hu* and 14 *bianzhong* (chime bell) set found within the same tomb at Xiaotianxi conform to the current fashion of northern central China during the late Warring States and early-Qin-to-Western Han periods. The inlay technique and design of vessels and bells can be compared with others excavated from Chu tombs of late Eastern Zhou date at Jiangling, or of Han date at Mawangdui in Hunan Province. The inlaid vessel and bell type, nonetheless, are bronze types popular throughout China during the Warring States period and should not be interpreted as objects unique to Chu.* Due to the traditional Ba paraphernalia of *zheng* and *chunyu,* the site is unequivocally Ba, not Chu under Ba influence. The drum and gong were used in Ba ritual performances and evidently were accoutrements reserved for the Ba elite.

Finally, the problems facing archaeology in the Three Gorges area are manifold and profound. If we are to understand the southern contribution to Chinese civilization, then archaeology must be supported in this area. To inundate the Three Gorges would not only eradicate the only source of our understanding but would also eradicate completely the contribution of the little understood but uniquely creative Ba peoples and their relationship to Shu and Chu.

Archaeological Sites," *Early China* 13 (1988): 116–152, and Wang Ran, "Ba Danshan, Danshui, Danyang yu Chu Digui zanyang bian," (Distinguishing the Ba Sites of Danshan, Danshui, and Danyang from the Chu Site of Zigui Danyang), *Changjiang wenhua lunji* 1 (1995): 267–275.

   *See, Lothar von Falkenhausen, *Suspended Music: Chime-Bells in the Culture of Bronze Age China* (Berkeley: University of California Press, 1993).

Appendix D

Priority-Level Cultural Antiquities in the Three Gorges Area

Total Number: 1,271

Above ground: 442 sites of four separate types:

1) Ancient structures: 229; 7 of which are priority sites*
2) Ancient bridges: 65
3) Stone carvings and sculptures: 117
4) Others: 31

Subsurface:** 829 sites, covering nine chronological periods:

1) Ancient bio-fossils: 14 sites
2) Paleolithic: 52
3) Neolithic: 85
4) Ba and Chu cultures: 150; 3 of which are priority sites that must be fully excavated
5) Qin-Han dynasties: 442
6) Six dynasties period: 31
7) Sui (A.D. 581–618) and Tang (A.D. 618–907) dynasties: 7
8) Song (A.D. 960–1279) and Yuan (A.D. 1271–1368) dynasties: 30
9) Ming (A.D. 1368–1644) and Qing (A.D. 1644–1911) dynasties: 18

Status: Officially approved national-level sites: 1
Provincial-level sites: 10
National-level sites awaiting approval: 8
Provincial-level sites awaiting approval: 50

---

*Includes architectural sites.
**Ruins (including tombs).

Appendix E

Archaeological Sites to Be Inundated in 1997 by the Construction of the Three Gorges Dam

1. Neolithic, Han dynasty, Six dynasties—Maozhaizi remains (*yizhi*); Badong County, Dongrangkou Town, Leijiaping Village
2. Neolithic, Han dynasty—Kongbaohe remains; Badong County, Dongrangkou Town, Jiaojiawan Village
3. Neolithic, Shang, Zhou, Han dynasties—Guandukou remains; Badong County, Guandukou Town, Dongpo Village
4. Shang, Zhou, Han, Tang dynasties—Xuetangbao remains; Badong County, Pingyangba Town, Longchuanhe Village
5. Neolithic, Han dynasty—Sijiping remains; Badong County, Wanliu Town, Sijiping Village
6. Neolithic—Huoyanshi remains; Badong County, Xinling Town, Huoyanshi Village
7. Neolithic, Zhou, Han, Tang dynasties—Guanzhuangping remains; Zigui County, Quyuan Town, Guanzhuangping Village
8. Neolithic, Shang, Zhou dynasties—Liulinxi remains; Zigui County, Maoping Town, Lanling agency (*banshichu*), Miaoke Village
9. Qing dynasty—Wangye temple; Badong County, Nanmuyuan Village, First group
10. Qing dynasty—"Chushuhonggou" stone engraving; Badong County, Guandukou Town, Wanliu port
11. Qing dynasty—*Lianzixi* stone engraving; Badong County, Yuntuo administrative district, Hongyanshi Village

12. Qing dynasty—*Lianzixi* towline site; Badong County, Yuntuo administrative district, Hongyanshi Village

13. Qing dynasty—Shuifu temple; Zigui County, Xiangxi Town, Xiangxi residence committee

14. Song dynasty—Yuxu cave, Moya stone carving; Zigui County, Xiangxi Town, Bazimen Village, Second group

15. Qing dynasty—Wangjia memorial; Yichang County, Taipingxi Town, Wuxiang temple

16. Southern Song dynasty—Suojiang pillar; Fengjie County, Qutang Gorge entrance, Tiezhu stream

Letter to Jiang Zemin Concerning Archaeological Sites, August 8, 1996

TO: Jiang Zemin, General Secretary of the Chinese Communist Party, and Comrades on the Standing Committee of the Politburo

Recently, we have become aware of the situation confronting historical relics and cultural antiquities in and around the site of the Three Gorges dam project. We feel it is our duty to bring this information to the attention of the highest authorities so that this matter can be attended to by the central government.

As the largest hydroelectric project in history, the Three Gorges dam has generated worldwide interest. This has included increased attention to the issue of cultural antiquities in and around the Three Gorges dam area. This area was the site of some of the earliest civilized developments in Chinese history and contains numerous subsurface and surface cultural antiquities. It is a prime example of the rich history and culture of the Chinese people. China is one of the most important places in the history of world civilization. The entire gamut of Paleolithic and Neolithic sites which have been unearthed in the vicinity of the Three Gorges area will allow researchers to evaluate the differences between eastern and western culture in the Yangtze River Valley as well as the relationship of these cultures with nature during ancient times. There was a group of people called the Ba who were known for their skills in making war and in the arts. They once helped King Wu of the Zhou eliminate the Shang dynasty, and they later established their own kingdom which was well-known throughout history but is now extinct. The Three Gorges area is the site of the Ba people and was their primary area of activity and development. The data uncovered so far by archaeological excavation will allow us to reconstruct the Ba Culture. But if not rescued in a timely fashion, the Ba will disappear with the construction of the reservoir—an irreversible error. In addition, various types of cultural antiquities associated with hydrology are evidence of our ancestors' attempts to utilize and control nature. Also, the long history of hydrology in the Three Gorges area is unique.

Especially important is the fact that the cultural antiquities in this area have long been tied to the area's beautiful natural scenery; the rich legacy of the area has been passed on by nature and our ancestors. It is the pride of the offspring of the Yellow Emperor. Such invaluable sites are admired at home and abroad as the jewels of Chinese civilization. For that reason, our government and engineering departments have on several occasions argued that we must do a good job in protecting these cultural antiquities so that our history and culture will continue to shine.

However, cultural antiquities preservation work is at present facing a serious challenge. As construction on the project has already begun, along with the initial stages of resettlement, some ancient graves and cultural antiquities have already been destroyed by giant earth-moving equipment. There have also been cases in which architecturally valuable sites have been destroyed in the process of resettlement. Recently, the carving known as "Conversing About Beautiful Mountains and Rivers" (*gonghua haoshanchuan*) in Badong County was dynamited. The destruction of this major relic epitomizes the dire situation confronting all cultural antiquities in the proposed reservoir area. We would especially like to note that due to the continued delay in funding the rescue effort for cultural antiquities in the Three Gorges area all efforts at rescuing the cultural antiquities have come to a halt, which will result in even more losses.

Another reason for this situation is that since the cultural antiquities departments were not directly involved in the original assessment of the Three Gorges project, it was left to other departments to determine the level of funding, based on the conditions of some 100 relic sites with which they were familiar. Based on that sparse information, these other departments also thought that the plan proposed by the cultural antiquities departments, and approved by resettlement departments, did not qualify the area as a "priority preservation site." But that conclusion was wrong. It is our understanding that the cultural antiquities departments once organized hundreds of professionals to conduct investigations and research in the reservoir area, where they discovered over twelve hundred such [important] sites. But due to the advanced stage in the launching of the dam project and the country's [perilous] financial situation, the design of the plan for cultural antiquities preservation singled out only a select number of sites to be unearthed, relocated, and/or preserved. Take for example the case of subsurface relics. According to data provided by cultural antiquities preservation planning departments, there are 829 known subsurface sites covering 20 million square meters. But areas designated for excavation amount to less than two million square meters, or less than one-tenth of the entire area. This indicates that the current plan reflects the policy of "preservation and excavation of priority sites." Thus, accusations that the cultural antiquities departments want to extend the scope of cultural antiquities preservation and are asking for too much money are groundless.

For this reason, we advocate and hope that the central government will speed up the mobilization of resources to assess and approve policies for cultural antiquities preservation and to provide adequate funding. As is well known, approval for such a policy is very time consuming. But the construction of the dam and resettlement continue unabated. Therefore, to prevent greater losses of cultural antiquities during construction [we ask] that before these plans for cultural antiquities preservation pass through the various stages of approval monies already allotted for cultural antiquities preservation be distributed so as to ease widespread concerns about the fate of these sites. The present situation is such that sections of the dam will be conjoined next year, causing the water level to rise to 82.28 meters. The river bank in Badong County, Hubei Province will be elevated to 105.4 meters. The water level farther away from the dam site will be even higher. Based on that increase, about 130 sites will be submerged. It is already a pressing task to rescue these sites within a year. But to date, the project departments have not provided any funding for this effort, something which has caused great consternation among people in the cultural antiquities departments.

In the forty-seven years since the establishment of the People's Republic of China, cultural antiquities work has made considerable progress. We have particularly excelled at protecting cultural antiquities on sites slated for capital construction. As an important part of socialist spiritual civilization, the profound significance of cultural antiquities preservation has been given more and more attention; this will play a greater role in maintaining the spiritual bond of the Chinese nation, strengthening the unity among the offspring of the Yellow Emperor both in China and abroad, and increasing national confidence. Thus today when the central government advocates doing a good job in the creation of a spiritual civilization, we feel that cultural antiquities preservation has even greater significance. We hope that cultural antiquities preservation in and around the Three Gorges project will be a starting point for integrating cultural antiquities preservation as part of patriotic education, increasing the understanding of the entire nation of the importance of cultural antiquities preservation, continuing the fine traditions of the Chinese nation, and strengthening the unity and inspiration of our great motherland.

As we march toward the twenty-first century and carry out modernization and achieve greater things, the burden and responsibility of preserving historical legacies and spreading the fine traditions of our culture are even greater.

Thus, at the moment when the Three Gorges cultural antiquities are about to be destroyed on a large scale, we respectfully present this letter to you in the hope that the Three Gorges project departments realize the importance of the cultural antiquities in the area and carry out rescue work jointly with cultural antiquities departments.

Signed in order:
   Su Bingqi, President of the Chinese Archaeological Society, and Research Fellow, Chinese Academy of Social Sciences, Archaeological Institute
   Zhang Kaiji, Master Architect and Chief Engineer, Beijing General Architectural Design Institute
   Zheng Xiaobian, Specialist in Urban Planning
   Wang Kun, Artist
   Luo Zhewen, Specialist in Ancient Architecture, and Deputy Director, China Great Wall Association
   Zhou Weizhi, Artist
   Zheng Siyuan, Director, China Cultural Antiquities Association
   Wang Meng, Writer [former Minister of Culture]
   Dan Shiyuan, Research Fellow, Palace Museum
   Chai Zemin, Director, China Diplomacy Association, and Deputy Director China Cultural Antiquities Association [former ambassador to the United States]
   Wang Dingguo, Deputy Director, China Cultural Antiquities Association
   Xie Chensheng, China Cultural Antiquities Association
   Lin Yaohua, Professor, Central University of Nationalities

Bing Xin, Writer

Ma Xueliang, Professor, Central University of Nationalities

Chen Yongling, Professor, Central University of Nationalities

Wang Zhonghan, Professor, Central University of Nationalities

Chen Liankai, Professor, Central University of Nationalities

Song Zhuohua, Professor, Central University of Nationalities

Zhuang Kongzhao, Professor, Central University of Nationalities

Fu Lianxing, Senior Engineer, Palace Museum

Zou Heng, Professor, Beijing University

Yan Wenming, Professor, Beijing University

Li Boqian, Professor, Beijing University

Su Bai, Professor, Beijing University

Wu Rongzeng, Professor, Beijing University

Li Jiahao, Professor, Beijing University

Qiu Xigui, Professor, Beijing University

Deng Guangming, Professor, Beijing University

Zhou Yiliang, Professor, Beijing University

Ji Meilin, Professor, Beijing University

Lü Zun'e, Professor, Beijing University

Wang Qiming, Professor, Beijing Architectural Engineering Institute

Jiang Zhongguang, Professor, Beijing Architectural Engineering Institute

Ye Zurun, Professor, Beijing Architectural Engineering Institute

Zang Erzhong, Professor, Beijing Architectural Engineering Institute

Zhang Zhongpei, Professor, Palace Museum

Xu Guangji, Research Fellow, Chinese Academy of Social Sciences, Archaeological Institute

Jia Lanpo, Research Fellow, Chinese Academy of Sciences, Institute of Ancient Vertebrae Animals and Ancient Anthropology

Hou Renzhi, Member, China Academy of Sciences, and Professor, Beijing University

Wu Liangyong, Member, Chinese Academy of Sciences and China Engineering Institute, and Professor, Qinghua University

Bai Shouyi, Professor, Beijing Normal University

Yu Weichao, Professor, National History Museum [Director, Cultural Antiquities Relocation, Three Gorges Project]

Tian Fang, Research Fellow, State Planning Commission, Institute of Economics

Bi Keguan, Research Fellow, China Arts Academy

Jin Yongtang, Professor, Hydrology and Hydroelectricity Academy

Huang Kezhong, Senior Engineer, China Institute of Cultural Antiquities

Huang Jinglü, Deputy Director, China Archaeological Society, and Research Fellow, China Institute of Cultural Antiquities

Tong Zhuzhen, Research Fellow, Chinese Academy of Social Sciences, Archaeological Institute

Wang Jin, Research Fellow, Hubei Province Cultural Antiquities Preservation Institute

Huang Zhanqiu, Research Fellow, Chinese Academy of Social Sciences, Archaeological Institute

Mu Xin, former Editor-in-Chief, *Enlightenment Daily* [*Guangming ribao*]

Wen Jizi, Professor, Chinese Academy of Social Sciences, Graduate School

Zhu Zixuan, Professor, Qinghua University

Bao Shixing, Secretary-in-Chief and Professor, China Urban Sciences Association

Du Baicao, Professor, China Architectural Technology Research Institute

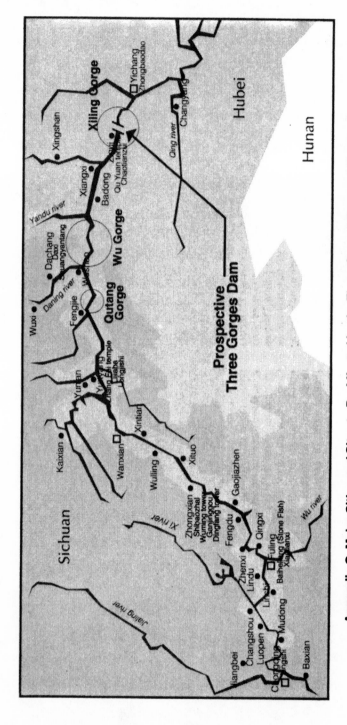

Appendix G. Major Cities and Sites to Be Affected by the Three Gorges Dam and Reservoir
(Adapted by Chris Ingersoll from *Zhongguo Changjiang Sanxia*, Hong Kong, 1993, pp. 26–27)

Biographic Glossary

Dai Qing: Born in 1941, Dai Qing was trained as a missile engineer at the Harbin Military Engineering Institute and is the adopted daughter of Ye Jianying, one of China's most senior military leaders. In the 1960s, Dai Qing became a Red Guard activist during the Cultural Revolution and underwent secret service training in the military. In the early 1980s, she became a journalist at *Guangming ribao* (Enlightenment Daily) where she began doing investigative reports on intellectual persecution throughout Chinese Communist Party (CCP) history. She reported on the cases of Wang Shiwei, Liang Shuming, and Chu Anping, all prominent intellectuals who were purged by Party leaders in the 1940s and the 1950s. Dai Qing is a strong advocate of press freedom and environmental protection, and she has collected documents from many scientists and economists opposed to the Three Gorges dam project. In 1989, Dai was imprisoned after the June fourth crackdown in Tiananmen Square. At the same time, her book on the Three Gorges dam, *Changjiang! Changjiang!* (Yangtze! Yangtze!), was banned for allegedly contributing to the "turmoil." Later released, Dai was allowed to travel abroad and has since been a Nieman Fellow at Harvard University and a fellow at the Freedom Forum, School of Journalism, Columbia University.

Deng Xiaoping: Born in 1904 in Sichuan Province, Deng Xiaoping was the eldest son of a landowner. In 1920, he traveled to France as a work-study student where he joined the European branch of the Chinese Communist Party. On returning to China in 1927, he assumed his first position as an instructor at the Xi'an Military and Political Academy. In 1929, he helped organize communist military forces in the southwestern province of Guangxi and became a political commissar. During the 1945–49 Civil War with the Nationalists, Deng was a member of the Second Field Army in the Crossing the Yangtze River and Huaihai battles. In 1952, he was appointed a vice premier and in 1956 became a member of the CCP Politburo Standing Committee and head of the Party Secretariat. He was condemned in the Cultural Revolution (1966–76) for having previously criticized the personality cult of Mao Zedong and for his "liberal" policies on agriculture and industry. He

221

reappeared in 1973 as a vice premier, and in 1975 he was reappointed to the Politburo Standing Committee, only to be dropped again in 1976 following the April Tiananmen demonstrations. Deng reappeared again in July 1977, when he assumed all of his previous posts and also became chief of staff of the People's Liberation Army. In 1981, he became chairman of the Central Military Commission. In November 1987, he "retired" from all posts except the Military Commission, a position which he finally yielded in November 1989. Deng Xiaoping died on February 19, 1997, at the age of ninety-two.

Guo Shuyan: Born in 1935 in Zhenping, Henan Province, Guo Shuyan joined the CCP in 1957. In 1959, he graduated with a degree in Metallurgy from the Polytechnical College of the Urals in the Soviet Union. In the same year, in the midst of the Great Leap Forward (1958–60), Guo returned to China and became deputy director of the Shenyang Manufacturing Research Institute of the First Ministry of Machine Building. He was also appointed deputy director and chief engineer of the High Energy Physics Research Institute of the Chinese Academy of Sciences, deputy director of the Bureau of Development Estimates of the State Science and Technology Commission, deputy director of the Science and Technology Leading Group of the State Council, deputy director of the State Science and Technology Commission, and vice governor and governor of Hubei Province. Guo also served as Party secretary of Hubei Province.

Huang Shunxing: Born in 1923 in Taiwan, Huang Shunxing was, from 1964 to 1969, a freely elected county executive in Taiwan. Huang subsequently moved to China and, since 1988, has been a member of the Standing Committee of the National People's Congress. He is an expert on agriculture and environmental protection.

Huang Wanli: Born in 1911 in Chuansha (now part of Shanghai) in Jiangsu Province, Huang Wanli graduated from the Department of Civil Engineering of Tangshan Communications University in 1932. He then traveled to the United States and in 1935 received a master's degree in civil engineering from Cornell University. In 1937, he earned a Ph.D. in the same field from the University of Illinois. Upon returning to China, under the Nationalist government, Huang became the technical supervisor of the Water Resources Department of the All-China Economic Committee, an engineer in the Water Resources Bureau of the Sichuan Province government, and engineer-in-chief and director of the Water Resources Bureau of Gansu Province in China's northwest. After the Communist takeover in 1949, Huang was appointed to professorships at Tangshan Railway College and Qinghua University where he conducted research in the field of water resources and hydrology (*shuiwenxue*) as well as water resources planning. He is the author of many works, including a major research study entitled "Harnessing the Yellow River."

Li Boning: A major figure in China's water resources bureaucracy, Li Boning is a primary supporter of the Three Gorges Dam Project. In the late 1950s, Li Boning was the deputy director of the Capital Construction Department of the Ministry of Water Resources and Electric Power. In 1978, he was identified as a vice minister of water resources and electric power. In 1989, he became the leading member of the Central Flood Prevention and Control Office. In 1988, he was elected vice chairman of the Economic Committee of the CPPCC.

Li Eding: Born in 1918 in the city of Tianjin, Li Eding graduated as a civil engineer from Qinghua University in 1940. In 1943, he moved to London and worked as a visiting scholar at an engineering consulting firm. In 1946, he returned to China and held several positions, including chief engineer and director of engineering in the Hydropower Department for the Longxi River, Changshou, of the Sichuan Water Resources Commission. After the Communist takeover in 1949, he became deputy director of the General Hydropower Bureau of the Ministry of Fuel Industry, deputy chief engineer of the Survey Design Research Institute of the Hydropower Bureau of the Ministry of Electric Power, and chief engineer of the Hydropower Engineering Bureau of the Shizitan dam project. In 1956, he became a member of the CCP, and in the same year he was recognized as a All-China Labor Hero. In subsequent years, he held many positions in the water resources bureaucracy, including: chief engineer in the Three Gate Gorge (*Sanmenxia*) Engineering Bureau and in the General Hydropower Bureau of the Liujia Gorge dam project; deputy chief engineer of the General Bureau of Hydropower Construction; chief engineer of the Capital Construction Bureau; senior engineer and vice minister of the Ministry of Water Resources and Electric Power; deputy chairman of the Association of Chinese Hydropower Engineering; chairman of the China Electrical Engineering Association; member of the Standing Committee of the China Water Resources Association; and deputy chairman of the International Dam Committee.

Li Fudu: Born in 1928 in Pucheng, Shaanxi Province, Li Fudu earned a Ph.D. from the Department of Water Resources of the Hanover (Germany) Polytechnical University. On returning to China, he was appointed chief engineer of the North China Water Resources Commission, director of the Water Resources Engineering Institute of Tianjin, director of the Research Office for Harnessing the Dan River of the Sichuan Water Resources Bureau, director of the Design Group of the Yellow River Planning Commission in Xi'an, and director of the Engineering Department of the same commission. After the Communist takeover in 1949, Li was appointed minister of water resources of the Northwest Military and Administrative Committee, deputy director of the Yellow River Planning Commission under the Ministry of Water Resources, and deputy chairman of the Henan Province Political Consultative Conference. He was also a member of the Standing Committee of the Sixth People's Congress of Henan Province, a member of the Central Committee of the Democratic Revolutionary Party (one of China's largely powerless satellite parties), a delegate to the second, third, fifth, and sixth National People's Congresses, and a member of the Chinese People's Political Consultative Conference. He is best known for his proposals for extensive reservoir construction and the buildup of sedimentation behind the Three Gate Gorge dam as solutions to the soil erosion problem on the middle reaches of the Yellow River.

Li Peng: Currently the premier of China, Li Peng is the adopted son of Zhou Enlai. Li Peng was born in 1928 in Sichuan Province to parents active in the CCP, both of whom were executed during the early 1930s. From 1948 to 1954, Li was trained as a power engineer in the Soviet Union, and from 1955 to 1979 he worked in China in numerous positions in the power industry. In 1982, he became vice minister of the Ministry of Water Conservancy and Power and in the same year became a member of the Central Committee at the Twelfth CCP Party Congress. In 1985, he was appointed to the Politburo and, in 1987, to its Standing Committee. He became premier in 1988. In June 1989, he reportedly transmitted the order issued by Deng Xiaoping for troops to use force against pro-democracy demonstrators. His official term as premier expires in March 1998.

Li Rui: Born in 1917, a one-time secretary on industrial affairs to Mao Zedong in the 1950s, Li Rui served as vice minister of electric power from 1955 to 1958, and in 1955 was appointed director of the General Bureau for Hydropower Construction in the Ministry of Electric Power. In 1956, Li Rui served on the Yellow River Planning Commission and the State Planning Commission. In 1958 and 1959, he was a vice minister of the Ministry of Water Resources and Electric Power. He was purged for his support of Peng Dehuai's opposition to the Great Leap Forward. In 1979, Li was rehabilitated and appointed vice minister of the power industry and director of the State Bureau of Computers. From 1982 to 1985, he was a member of the CCP Central Committee, and in 1985, he was appointed to the Central Advisory Commission, a largely honorific organization established for semiretired Party leaders.

Li Xiannian: Born in 1909 to poor peasants in Hubei, Li Xiannian was trained as a carpenter and then joined the Communists in 1927. Rising to the top of the CCP hierarchy as a military commander, Li Xiannian became a member of the Central Committee in 1945, and after 1949 he became head of Wuhan and in 1954 Party Secretary of Hubei Province and a vice premier. He was made a member of the CCP Politburo in 1956 and became minister of finance in 1957. He continued to serve on the Politburo throughout the Cultural Revolution (1966–76) and remained a central figure in economic and financial affairs through the 1980s. From 1983 to 1988, he was president of the People's Republic of China. Li Xiannian died in 1992.

Liu Jianxun: A veteran of the Shanxi Province "Dare-to-Die" corps in the 1930s, Liu Jianxun became the first secretary of Guangxi Province in 1957, and served from 1958 to 1961 in the same post in the reconstituted Guangxi Autonomous Region. In 1958, he also became an alternate member of the CCP Central Committee. He served as first secretary of Henan Province from 1961 to 1966 and from 1971 to 1978. Liu Jianxun disappeared in 1978 due to "grave errors and crimes," and in 1980 he lost his last remaining post as a delegate from Henan Province to the National People's Congress.

Liu Lanpo: As vice minister of the Ministry of Fuel Industry in 1954, Liu Lanpo accompanied Li Rui to the Soviet Union where he met Li Peng, who was studying to become a hydrological engineer. During the ensuing years,

Liu Lanpo had Li Peng appointed to positions in the Chinese electric power bureaucracy so as to groom him for future leadership. Purged during the Cultural Revolution, Liu was rehabilitated in 1979 and appointed minister of power industry, where he appointed Li Peng as his successor.

Mao Zedong: Chairman of the CCP from 1938 until his death in September 1976.

Peng Dehuai: Born in 1898 in Hunan Province, Peng Dehuai left his family at an early age and joined local military forces. In 1919 he was profoundly influenced by the writings of Sun Yat-sen and the liberal ideas of the 1919–25 May Fourth Movement. Peng joined the CCP in 1928 and became one of the foremost military figures in the communist movement, commanding CCP forces in a major battle with the Japanese in 1940 and leading the First Field Army during the Civil War with the Nationalists. Peng then commanded Chinese forces during the 1950–53 Korean War in which his troops fought to a standstill with American forces, but with extremely heavy losses on the Chinese side. A strong supporter of a professional military in China, Peng helped introduce ranks in 1954 and he became a marshal, the highest ranking position in the PLA, as well as minister of national defense. Peng's letter, in August 1959, to Mao Zedong raising questions about economic policy in the Great Leap led to his purge. Efforts to rehabilitate Peng in the early 1960s provoked Mao's wrath, and during the Cultural Revolution Peng was denounced and paraded through the streets by Red Guards. Peng died in obscurity in 1974.

Qian Zhengying: Born in 1923 in the United States and trained as a civil engineer in China, Qian Zhengying was appointed deputy director of the Water Resources Department in the East China Military and Administrative Council in 1950. In 1952, she became vice minister of water resources, and in 1957 she was made vice chairman of the Commission for Harnessing the Huai River. In 1958, she became vice minister of the new joint Ministry of Water Resources and Electric Power. In 1975, Qian was appointed minister of water resources and electric power and, in 1983, she visited the United States as part of an electric power delegation. In 1982, she was appointed to the CCP Central Committee and reappointed in 1987. Since the mid-1980s, Qian Zhengying has strongly supported construction of the Three Gorges Dam.

Qiao Shi: Leader of the Shanghai student movement in the 1940s, Qiao Shi worked in the Communist Youth League and in the steel industry in the 1950s and early 1960s. In 1982, he was appointed director of the International Liaison Department of the CCP, and in 1984, he was appointed director of the Party Organization Department. In 1985, he became a member of the Politburo and the Party Secretariat specializing in political and legal work and became head of the Political and Legal Affairs Commission of the CCP. In 1987, he was appointed to the Politburo Standing Committee, and since 1993 he has headed the National People's Congress.

Tan Zhenlin: A early follower of Mao Zedong in the 1927 Autumn Harvest uprising, Tan Zhenlin became a political commissar in the New Fourth Army. In 1956, he was appointed to the Secretariat of the Central Committee and in 1962 became a vice chairman of the State Planning Commission. In 1967, he participated in the so-called "February Adverse Current" that attempted to terminate the Cultural Revolution and outlaw the Red Guards. He was purged in 1967 and reappeared in 1973, when he was reappointed to the Central Committee.

Wang Huayun: Born in Guangtao, Zhili (today's Hebei Province), in 1908, Wang Huayun graduated from the Department of Law of Beijing University in 1935. In 1938, he joined the CCP and held numerous posts in the Chinese Communist border region government, including director of the Yellow River Planning Commission of the Ji-Lu-Yü (Hebei, Shandong, Henan) Border Region. After the communist takeover in 1949, he was appointed vice minister of the Ministry of Water Resources and director of the Yellow River Planning Commission, where he became involved in the campaign to harness the Yellow River through numerous proposals involving the Three Gate Gorge dam.

Wen Shanzhang: Trained as a hydrologist in the Soviet Union, Wen Shanzhang has served as a senior engineer of the Yellow River Water Resources Planning Research Institute (Zhongguo Huanghe shuili guihua yanjiuyuan).

Xi Zhongxun: A political commissar in the Northwest PLA during the Civil War and a director of the Communist Party Propaganda Department, Xi Zhongxun was elected to the Central Committee of the CCP in

1956. In 1962, he disappeared as a result of his close association with Peng Dehuai. In 1978, he reappeared as Party secretary of Guangdong Province and Second Political Commissar of the Guangzhou Military Region. He was a member of the Politburo from 1982 to 1987.

Zhang Hanying: Trained as a civil engineer at the University of Illinois and as a hydrologist at Cornell University, Zhang Hanying was appointed chief engineer to the Yellow River Planning Commission in 1933, during the Nationalist era. After the Communist takeover in 1949, Zhang was appointed vice minister of the Ministry of Water Resources specializing in dam design and surveying.

Zhou Enlai: Perhaps the most astute and cosmopolitan politician among top CCP leaders, Zhou Enlai was born in 1898 to a well-to-do gentry family. After 1949, Zhou served as foreign minister and premier. He endorsed a liberalization of policies toward intellectuals in the 1950s and assumed initially a neutral position on the Great Leap Forward. Zhou stuck with Mao Zedong through thick and thin, and during the Cultural Revolution Zhou reluctantly supported the widespread purges of Party leaders but made every effort to protect old colleagues from Red Guard attacks. Zhou also provided protection for China's historical relics, such as the Forbidden City, which were often targeted for destruction by the rampaging Red Guards. Zhou Enlai died in January 1976.

Zou Jiahua: A graduate of a Moscow Engineering Institute, Zou Jiahua served in the 1950s and 1960s as a director of a machine tool plant in Shenyang City in Northeast China, and then he worked in the First Ministry of Machine Building. In 1977, he was identified as a deputy director of the National Defense Industry Office under the State Council and he was elected as an alternate member of the CCP Central Committee. In 1982, he became vice minister of the Commission of Science, Technology and Industry for National Defense, and he was appointed minister of ordnance industry. In 1988, he became a state councillor and minister of machine building and the electronics industry. In 1991, he became head of the State Planning Commission and a vice premier. He became a member of the Politburo at the Fourteenth Party Congress in 1992.

Index

Agriculture:
  cash crops and, 48, 87, 180
  dry-land skills, 80–81
  grain output, 60, 61
  grain production, 10
  loss of, 4
  new lands for, 47, 59, 66
  post Cultural Revolution, 8
  specialized products, 69
  Sputnik communes and, 110–111
  tractors use of, 81
  See also Cash crops; Farmers;
    Macroagriculture
Agriculture Responsibility System,
    82, 100
Anhui Province, 30
  flood in, 26
  river irrigation network in, 20
Anti-Confucius Campaign, 117
Anti-Rightist struggle, 112
Antiquities Association, 214–219
Archaeological Institute of Sichuan,
    206
Archaeology:
  above ground sites, 134
  agencies for, 125, 140, 141
  ancient cultures and, 128–130
  below ground sites, 127–130
  Cultural Antiquities Law, 130–131
  dam height and, 135
  folk culture, 134–135
  funding for, 131, 136, 138–139, 189
  international cooperation, 135, 140–
    141
  location of, 130

Archaeology *(continued)*
  ranking of cultural antiquities,
    130–131
  relics types of, 126
  site locations of, 133, 212–213
  staff needed for, 133
  temples and, 126–127
  time for, 133–134
  vs resettlement budget, 132–133
"August 1975 Disaster":
  Banqiao Dam collapse, 33
  dynamite dams in, 36
  events after, 33–34
  events prior, 32–33
  impact of, 34–35
  Ji Dengkui and, 35–36
  Qian Zhengying and, 36–38
  Shimantan Dam collapses, 33
  Three Gorges Project and, 36
Autocratic system:
  closed-door decision making, 12–14
  local leader appointments in, 13
  Three Gorges Project and, 12
  undemocratic procedures in, 14
  vs communism, 12
  vs open system, 14
  See also Communist System;
    National People's Congress

Ba Culture, 129–130, 204–210
  art works of, 206–207
  artifacts of, 210
  Three Gorges and, 210
  war drums of, 207–208
  weaponry of, 208–209

229